AUTUMN HAWK FLIGHTS

Donald S. Heintzelman

AUTUMN HAWK FLIGHTS

*The Migrations in
Eastern North America*

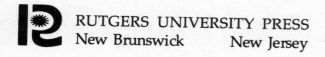

RUTGERS UNIVERSITY PRESS
New Brunswick   New Jersey

49790

Library of Congress Cataloging in Publication Data

Heintzelman, Donald S

    Autumn hawk flights.

    Bibliography: p.
    1. Hawks. 2. Birds—Migration. 3. Bird watching.
4. Birds—North America. I. Title.
QL696.F3H44   598.9'1   74-13918
ISBN 0-8135-0777-4

To the memory of my father,
who always enjoyed hearing about each day's hawk flight

Contents

PART 2. THE HAWK LOOKOUTS

Illustrations

PHOTOGRAPHS

FIGURES

MAPS

Tables

Preface

For the past eighteen years it has been my good fortune to be able to study the spectacular autumn hawk migrations which occur in eastern North America. During this period, I became aware that a voluminous amount of information existed regarding the subject. However, much of it was scattered in the periodical literature. A sizeable quantity of data also remained unpublished. The raptor bibliography developed by Olendorff and Olendorff (1968, 1969, 1970) was helpful in guiding me to some of the periodical literature.

Although information dealing with autumn hawk migrations in eastern North America now is voluminous, no synthesis of the type attempted here has been published previously. Even Jean Dorst (1961), in his splendid volume *The Migrations of Birds,* provided only a brief discussion. More recently, two overviews of portions of the subject appeared: *A Guide to Northeastern Hawk Watching* (Heintzelman, 1972d) and *A Study of Hawk Migration in Eastern North America* (Haugh, 1972). Bent (1937, 1938) also summarized migration dates for North American raptors, and a detailed account of the migrations at Hawk Mountain, Pennsylvania, was presented by Maurice Broun (1949a) in *Hawks Aloft: The Story of Hawk Mountain.* Additional details regarding the migrations at Hawk Mountain were published in *Feathers in the Wind* (Brett and Nagy, 1973) and in *The View from Hawk Mountain* (Harwood, 1973).

   This book focuses particular attention upon the hawk migrations in Pennsylvania, New Jersey, and other Middle Atlantic states.* However, it also deals with migrations crossing New England and eastern Canada, the southern Appalachian Mountain states, the southern coastal and Gulf coast states, and the Great Lakes states and Canadian provinces. A chapter also is devoted to autumn migrations of Broad-winged Hawks in Central America. Although most of the text is descriptive, I have made some preliminary analyses of certain phenomena. Whenever possible I have avoided use of technical terms to make the book useful to as many people as possible. The appendix lists the scientific names of all species mentioned. Literature appearing after 1971 was cited only if it was of exceptional importance, if it contained data from 1971 or earlier, or was available readily without a special search being made.

   The book is divided into six parts. Part 1 describes field study methods and discusses hawk identification. Part 2 provides an extensive survey of the most important or useful eastern North American hawk lookouts. The survey thus serves as a catalog of sites of exceptional ornithological importance. Hopefully conservationists will be able to preserve the most important of these. Birders also will be able to benefit from this survey by being able to visit many more hawk lookouts than previously were known to them. Parts 3 through 6 detail more technical aspects of autumn hawk migrations. Even novices will find much of this information of value, however. The list of literature cited also will lead interested persons to sources of much fascinating reading.

   My hawk migration research began in 1957 and continued during the period 1961 through 1973 at Bake Oven Knob, Pennsylvania. These field studies, conducted with the cooperation and/or assistance of other qualified observers, numbered 799 days. A total of 143,375 vultures, hawks, eagles, harriers, Ospreys, and falcons were

---

* Spring hawk migrations still are studied relatively little. These flights tend to disperse, rather than to concentrate at many geographic locations, as they move northward. However, the south shorelines of Lakes Erie and Ontario are major concentration points. Recent interest in spring migrations should provide information leading to a better understanding of these flights. Haugh and Cade (1966) and Haugh (1972) have made the most detailed studies of spring hawk flights to date.

counted. I relied heavily upon this large quantity of data in developing various concepts presented here. Published and unpublished data from other lookouts enhance and supplement the Bake Oven Knob information. In point of fact, I made every possible effort to locate pertinent information dealing with autumn hawk migrations in eastern North America.

Despite the substantial quantity of information presented in this book, innumerable questions still remain unanswered regarding autumn (and spring) hawk migrations. This whole subject offers exceptional opportunities for amateur and professional hawk watchers to add to our basic fund of ornithological knowledge. There is no end to the variety of questions demanding answers. They range from basic questions of methodology and field study techniques to highly theoretical questions dealing with subjects such as the evolution and development of hawk migration behavior patterns. For example, can radar help to solve the existing questions regarding the possible movements of Broad-winged Hawks over large bodies of water? Similarly, can sailplanes or other types of aircraft be helpful in tracking migrating hawks moving along mountain ridges or over open country? What are the geographic origins of the hawks observed at the various lookouts? Do birds from a particular breeding ground use the same migration route yearly, thus appearing at the same lookouts each autumn? Or do hawk migration routes change yearly, or from time to time, or at random? What are the food requirements of migrating hawks? Can counts of migrating hawks be used as valid indices to raptor population trends? Perhaps this book will help to interest other persons in the study of hawk migrations and, in turn, inspire others to help to search for answers to some of these questions.

During the years of research leading to the preparation of this volume a large number of people assisted in countless ways. It is not practical to name each one, but I should like to call attention to those individuals whose assistance was of more than minor importance.

D. L. Knohr of Enola, Pennsylvania, was particularly helpful in providing and allowing me to use nearly two decades of previously unpublished hawk counts from Waggoner's Gap, Pennsylvania. Carl L. Garner also provided several years of unpublished data from his lookout at The Pulpit on Tuscarora Mountain, Pennsylvania. Fred

Scott sent me a copy of William Rusling's difficult to obtain unpublished manuscript detailing the 1936 hawk flights at Cape Charles, Virginia, and Richard H. Pough (who financed that study) granted me permission to publish excerpts from it.

A special note of appreciation also is due Chandler S. Robbins of the United States Fish and Wildlife Service, who permitted me to make use of the extensive file of unpublished hawk migration data preserved under his care at the Patuxent Wildlife Research Center, Laurel, Maryland. He also arranged for my unrestricted access to the Center's research library, which greatly facilitated the completion of my search of the literature.

Other persons who provided valuable information or assistance include D. Amadon, A. H. Bergey, A. Bihun, Jr., I. Black, J. Bond, A. Brady, F. Brock, E. A. Choate, G. B. David II, H. E. Douglas, H. Drinkwater, E. Eisenmann, T. W. Finucane, F. T. Fitzpatrick, E. W. Graham, A. Grout, P. Grout, R. L. Haines, R. R. Hendrick, B. C. Hiatt, C. M. Hoff, P. B. Hofslund, J. B. Holt, Jr., D. Hopkins, J. A. Jacobs, J. W. Key, B. Lake, C. L. Leck, W. Lent, G. H. Lowery, Jr., F. Mears, E. O. Mills, I. Morrin, R. Moser, B. G. Murray, Jr., A. Nagy, O. S. Pettingill, Jr., E. L. Poole, R. H. Pough, C. J. Robertson, J. L. Ruos, A. A. Sexauer, R. W. Smart, V. Smiley, R. W. Smith, D. Steffey, G. Steffey, S. Thomas, F. Tilly, J. Tobias, W. C. Townsend, A. Webster, C. Wellman, C. E. Wonderly, E. Wonderly, and K. Zindle.

Appreciation also is extended to the editors of the following journals for permitting me to quote lengthy excerpts, and/or reproduce illustrations, from material originally appearing in their publications:

Atlantic Naturalist for excerpts from "Hawk Watch" by C. S. Robbins (1956) and "Bake Oven Knob Hawk Flights" by D. S. Heintzelman (1963).

The Auk for excerpts from "Migration of Hawks" by J. von Lengerke (1908), "Hawk Notes from Sterrett's Gap, Pennsylvania" by E. S. Frey (1940), and "The Daily Rhythm of Hawk Migration at Cedar Grove, Wisconsin" by H. C. Mueller and D. D. Berger (1973) and various illustrations.

Audubon Field Notes (now *American Birds*) for excerpts from the "Florida Region" report by W. B. Robertson, Jr., and J. C. Ogden (1968).

Audubon Magazine (now *Audubon*, the magazine of the National Audubon Society) for excerpts from "Hawk Migrations Around the Great Lakes" by O. S. Pettingill, Jr. (1962).

Bird-Banding for excerpts from "Autumnal Hawk Migration Through Panama" by D. L. Hicks, D. T. Rogers, Jr., and G. I. Child (1966).

Cassinia for excerpts from "An Intensive Inter-specific Aerial Display between a Sharp-shinned Hawk and a Sparrow Hawk" by D. S. Heintzelman (1970).

Delaware Conservationist for excerpts from "Our Handsome Birds of Prey: Hawks and Eagles" by C. E. Mohr (1969) and various illustrations.

EBBA News for excerpts from "Banding North American Migrants in Panama" by V. M. Kleen (1969).

Kingbird for excerpts from "Fall Hawk Watch at Mt. Peter" by S. F. Bailey (1967).

Scientific American for illustrations from "The Soaring Flight of Birds" by Clarence D. Cone, Jr. (1962).

Search (published by the Cornell University Agricultural Experiment Station, New York State College of Agriculture and Life Sciences, a Statutory College of the State University, Cornell University, Ithaca, New York) for excerpts and illustrations from "A Study of Hawk Migration in Eastern North America" by J. R. Haugh (1972).

Urner Field Observer for excerpts from "Blawking: The Study of Hawks in Flight from a Blimp" by E. I. Stearns (1948).

Wilson Bulletin for excerpts from "The Study of Hawks in Flight from a Blimp" by E. I. Stearns (1949) and various illustrations.

Appreciation also is extended to the authors and publishers for brief quotations from various other works included in the list of literature cited.

B. C. Hiatt, C. M. Hoff, and C. F. Leck also kindly provided lengthy unpublished material, portions of which are quoted.

Fred Tilly deserves a special note of thanks for providing numerous outstanding photographs used as illustrations in this book. Other persons or organizations who also provided valued photographs include Acadia National Park, Dorothy Beatty, Jack Blount, Alan Brady, Harry Goldman, Karl L. Maslowski, the Newark Museum, Elizabeth W. Phinney, Steve Piper, Richard H. Pough, Niel G. Smith, Walter R. Spofford, Edwin I. Stearns, Tommy Swindell, and the U.S. Forest Service.

Portions of my field studies, reports on which already have been published, were conducted while I held positions as Associate Curator of Natural Science and Curator of Ornithology at the William Penn Memorial Museum and the New Jersey State Museum.

Finally, I should like to call very special attention to Robert and Anne MacClay of Cressona, Pennsylvania. Were it not for their extremely helpful assistance at Bake Oven Knob, and their constant encouragement, I should long ago have abandoned this project. Yearly, they willingly devoted many weeks of precious vacation time in making hawk counts at the Knob thus providing additional reference data essential to the completion of this book.

Allentown, Pa.   Donald S. Heintzelman
3 December 1973

Part 1
Field Study Methods
and Hawk Identification

1

Introduction

The first person in the Americas to record information on hawk migrations was the naturalist Oviedo. During the period 1526 to 1535, he noted hawks migrating across the West Indies and presented his data in *Historia general y natural de las Indias, islas y tierre-firme del mar oceano* (Baughman, 1947). Two centuries passed before another reference to hawk migrations was published in the New World. Then, in 1756, an account of what was probably a very large flight of Broad-winged Hawks over Brattleboro, Vermont, on 20 or 21 September, appeared in the *New Hampshire Gazette* for 14 October 1756 (Goldman, 1970). In general, however, it is only recently that naturalists and ornithologists have developed an awareness and special interest in the spectacular autumn migrations of hawks in eastern North America. A good deal of this interest originated as an outgrowth of the concern exhibited by conservationists dedicated to the preservation of North American birds of prey.

During recent years, thousands of birders in particular have discovered the excitement of watching migrating hawks. When someone visits a hawk lookout, however, it is likely that he will experience far more than just an adventure in superb hawk watching. Many of these sites are located on major migration routes used by

This Red-tailed Hawk, shot illegally during migration, suffered a shattered wing and was kept in captivity for many years at Hawk Mountain Sanctuary. Photo by Donald S. Heintzelman.

a wide assortment of other birds. Indeed, vast numbers of waterfowl and songbirds also migrate southward along the same coastlines, mountain ridges, and lake shorelines which are important migration routes for hawks. Hence splendid birding opportunities of all types exist at many hawk lookouts. Part of the adventure of going to Hawk Mountain, Pennsylvania, is that one never knows what might appear unexpectedly. For example, more than 232 species of birds have been identified within the Sanctuary since its inception in 1934 (Brett and Nagy, 1973). Among the most extraordinary of these avian visitors was a petrel from the South Pacific—a waif carried there by the northward movements of a hurricane (Heintzelman, 1961). Similarly, at nearby Bake Oven Knob more than 118 species

of birds have been identified during autumn since 1961 (Heintzelman, 1969b), and the number increases yearly as field studies continue. During one season, for example, an extraordinary migration
of White-breasted Nuthatches occurred and was as exciting as the
hawk migrations themselves (Heintzelman and MacClay, 1971).
Hence a visit to a hawk lookout always is a very special experience.

FIELD STUDY METHODS

There are many methods applicable to the study of hawk migrations. Some have been standard ornithological techniques for decades. These are described by Heintzelman (1972d) and Haugh
(1972). Other methods employ expensive and/or complex electronic
equipment not readily available to most hawk watchers. Only the
traditional methods were used at Bake Oven Knob.

VISUAL OBSERVATIONS

Visual observation is the standard method used for studying
hawk migrations. One stations himself on a favorable lookout on a
flyway and scans the sky and horizon for approaching hawks.

During the Broad-wing season, in September, observations usually begin at 0700. However, on days immediately following a major flight, or when a major flight is expected, observations often begin as early as 0630 at Bake Oven Knob. This is shortly before the
first thermals * of the day develop.

September hawk flights are the most prolonged of the season
in respect to hours of duration per day. Broad-winged Hawks usually stop flying only when thermal activity is severely reduced in
intensity. Then we often see hawks approaching head-on only to
watch them plunge into the surrounding forest for the night. Thus
it is often necessary to continue observations until 1800 hours or
later. On the other hand, Bald Eagles and Ospreys, which are much
less dependent upon thermals, usually remain aloft long after Broad-
wings land. These late afternoon flights are majestic. The birds appear in mellow light, often close to the lookout. The extra time spent

* Bubbles of warm air rising into the atmosphere. Chapter 14 provides details
regarding their formation and their use by migrating hawks.

Upper, flocks or "kettles" of Broad-winged Hawks milling inside thermals high-light the mid-September migration season. Photo by Alan Brady. *Lower left,* Sharp-shinned Hawks are the most abundant migrating raptors during early to mid-October. Photo by Donald S. Heintzelman. *Lower right,* Red-tailed Hawks are seen in largest numbers during late October and early November. Photo by Fred Tilly.

watching eagles and Ospreys gliding by on set wings is not only very enjoyable but also a necessary part of making accurate hawk counts.

October hawk flights differ from September flights in several respects. To begin, the bulk of the Broad-wings have passed. A far greater variety of raptors comprise the October flights. Sharp-shinned Hawks account for the bulk of the flights during early October. During the latter part of the month Red-tailed Hawks, Golden Eagles, and other large raptors appear in greater numbers. In addition, October hawk flights last for fewer hours per day because of the shorter days. At Bake Oven Knob our observations usually begin at about 0800 hours and continue until about 1730.

November hawk flights are still more time-compressed. Early in the month large Red-tailed Hawk flights occur. Goshawks, Red-shouldered Hawks, Rough-legged Hawks, and Golden Eagles frequently are included. However, by the latter half of the month and into early December, only a trickle of hawks and a few eagles are still migrating southward. Observations during November usually begin at about 0900 hours and continue to about 1600.

BINOCULARS AND TELESCOPES

A good pair of binoculars and a telescope are essential pieces of field equipment for studying hawk migrations. Generally 10× center-focus binoculars are ideal. Sometimes even 10× binoculars are inadequate, however. This is particularly true when hawks are flying at very high altitude, at a distance over valleys, or under dim light conditions. Then a 20× telescope is necessary to make positive identifications. For maneuverability, it is best mounted on a gun-stock.

The correct use of these instruments also is important. Experienced observers usually scan the skies and the horizon with binoculars at periodic intervals to detect distant flying hawks which otherwise might pass unseen and uncounted. But during the Broad-wing season, when kettles * of these birds rise to extremely high altitudes and are difficult to detect against blue skies, a scan of the sky and horizon with a telescope sometimes produces additional birds.

* A group or flock of hawks milling around inside a thermal.

Binoculars are essential for studying hawk migrations. Many experienced observers prefer 10× instruments. Photo by Donald S. Heintzelman.

A telescope, mounted on a gunstock, is useful for identifying hawks flying at a distance or under dim light conditions. Photo by Donald S. Heintzelman.

HAWK MIGRATION DATA SHEET

Date: _____ Observers: _____

Location: _____

Time (E.S.T.)	7-8	8-9	9-10	10-11	11-12	12-1	1-2	2-3	3-4	4-5	5-6	6-7	Totals
Max. Vis. (Miles)													
Air Temp.													
Wind Speed (MPH)													
Wind Direction													
% Cloud Cover													
Turkey Vulture													
Goshawk													
Sharp-shinned Hawk													
Cooper's Hawk													
Red-tailed Hawk													
Red-shouldered Hawk													
Broad-winged Hawk													
Rough-legged Hawk													
Golden Eagle													
Bald Eagle													
Marsh Hawk													
Osprey													
Peregrine Falcon													
Pigeon Hawk													
Sparrow Hawk													
Unidentified Hawk													
Totals													

NOTES:

FIGURE 1   The form used for recording data on migrating hawks at Bake Oven Knob, Pa. Similar forms are used at Hook Mountain and Mount Peter, N.Y.

FIELD DATA FORMS

Most hawk watchers now keep hourly counts of the various species passing their lookouts. Printed data forms greatly facilitate this record keeping. In addition, it is also desirable to record the exact times (Eastern Standard Time) when rare birds such as Golden Eagles, Bald Eagles, and Peregrine Falcons appear. For species such as the Rough-legged Hawk, the color phase * of the bird also should be recorded. Such detailed records frequently are helpful in tracking movements of individual birds over limited geographic areas. Estimates of the ages of endangered species are particularly important and should be carefully noted. During recent years, observers at Hawk Mountain Sanctuary, Pennsylvania, have recorded the ages (adult, immature, or undetermined) of as many hawks as possible.

* Some species are dimorphic, occurring as dark or light individuals. These differences are called color phases. Other types of dimorphism also occur in various species, e.g., among hawks, females are about one-third larger than males.

Report by: _____ (a.m.)   MONTCLAIR BIRD CLUB - DAILY HAWK COUNT   Date:_____
                               (p.m.)

Time	BW	Buteos	RT	RS	Accipiters SS	CH	Falcons Sp H	DH	PH	OS	Other MH	BE	Un	Total
-9														
9-10														
10-11														
11-12														
12-1														
1-2														
2-3														
3-4														
4-5														
5+														
Total														

FIGURE 2   The form used for recording data on migrating hawks at the Montclair Hawk Lookout Sanctuary, N.J. Courtesy of Andrew Bihun, Jr., and the Montclair Bird Club.

ESTIMATING WIND VELOCITY

To make accurate measurements of wind velocity requires expensive and complex equipment. Haugh (1972), for example, used a Windscope to determine wind velocity in his hawk migration study. At most lookouts such equipment is not available. Nonetheless, reasonably accurate hourly measurements of wind velocity can be estimated. Some observers allow a cloth to flutter in the breeze and estimate velocity from this. With a little experience, reasonable approximations can be made. Plastic wind speed gauges are more accurate. They may be purchased from camping equipment supply stores. Use of the Beaufort Scale of Wind Force is still another method of estimating wind velocity.

RECORDING WIND DIRECTION

Wind direction is an extremely important factor affecting autumn hawk migrations. It is important that hourly records be maintained. Haugh (1972) used his Windscope to determine surface wind directions. At Bake Oven Knob, and most other lookouts, a cloth is used as a wind sock, and its movement is compared against known compass directions. Either method determines surface wind direction only.

It is more difficult to determine wind directions at higher altitudes. Edwards (1939) devised a technique to solve this problem by filling small balloons with gas (helium) and releasing them into the atmosphere. By watching their movements as they rose, an approximation of higher altitude wind directions was secured. In 1967, I experimented with this technique at Bake Oven Knob. Helium-filled balloons were released hourly. In most cases surface winds were roughly equal to higher altitude wind directions. In 1968, hawk watchers at Hawk Mountain used this technique. Haugh (1972) also used a similar technique.

RECORDING AIR TEMPERATURES

Air temperatures also are components of the weather data which hawk watchers should develop for their lookouts. Any accurate thermometer can be used. At Bake Oven Knob I use an instrument calibrated in degrees Centigrade. The thermometer is suspended from a branch or limb which permits a free flow of air to circulate around it. Hourly air temperatures are entered onto a field data sheet.

ESTIMATING VISIBILITY

The degree of visibility at a lookout unquestionably affects the numbers of hawks counted. On clear days one can expect to *see* most hawks passing a lookout. But when extreme haze or air pollution occurs, many birds can pass uncounted. Hence estimates of visibility are desirable. They, too, are recorded hourly on data sheets.

ESTIMATING PERCENTAGE OF CLOUD COVER

Hourly estimates of the percentage of cloud cover over a lookout can help to reveal important aspects of the mechanics of hawk migrations. My studies of migrating Broad-winged Hawks at Bake Oven Knob, for example, reveal that estimates of the percentage of cloud cover are essential to understanding the migrations of these birds. Clouds are important factors which help to regulate thermal activity, and Broad-wings are highly dependent upon thermals. Although Haugh (1972: 3) considered cloud cover relatively unimportant to migrating hawks, in point of fact relatively little is known about the movements and navigational abilities of birds in opaque clouds (Griffin, 1969, 1973). Anyone observing hawks flying above or through clouds should make very detailed notes on the behavior of the birds. Such information may provide important insights leading toward a better understanding of this aspect of bird migration and navigation.

HAWK-COUNTING TECHNIQUES

In most instances, migrating hawks are easy to count because they appear individually or in small groups. However, when hundreds of Broad-winged Hawks form kettles in thermals it is difficult or impossible to make accurate counts. Experienced observers then wait until the hawks glide overhead in bomberlike formations, after the thermals dissipate, before counting them. Since thousands of hawks sometimes pass within relatively brief periods of time, a hand tally facilitates the counting process. The observer rapidly clicks off each bird as it passes overhead. If a flight continues into a new hour period, another tally counter is used for the hawks appearing in the new time. After the flight is completed, data for the previous hour (recorded on the first tally counter) are recorded on field data forms. Hand tally counters are used only for a single species (the most abundant migrant at the time).

On rare occasions, very large numbers of hawks pass a lookout during such brief periods of time that it is impossible to count them. One then is forced to estimate the numbers of birds seen. Broun (1949a: 185–186) described such a flight on 16 September 1948 at Hawk Mountain, Pennsylvania. More than 11,392 hawks (mainly Broad-wings) were reported. The extraordinary Sparrow Hawk flight on 16 October 1970 at Cape May Point, New Jersey, is another example. Choate (1972) estimated about 25,000 individuals passed

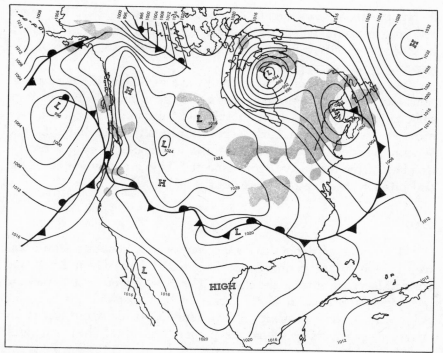

MAP 1 Weather maps, which show the positions of low- and high-pressure areas, cold fronts, and other weather features, are useful in predicting and studying autumn hawk migrations.

the Point. Estimating the numbers of hawks passing a lookout should be done only when it is impossible to make actual counts.

WEATHER MAPS

Weather systems play important roles in autumn hawk migrations either by acting as releasers,* thereby causing hawks to begin migrating (Robbins, 1956), or by affecting their migrations once the movements have begun (Broun, 1949a; Haugh, 1972; Heintzelman and Armentano, 1964; Mueller and Berger, 1961). Thus many hawk watchers have found weather maps useful as aids in helping

---

* External stimuli which cause birds to begin migrating. Changes in air temperature and/or barometric pressure frequently act as releasers.

to predict good hawk flights (Broun, 1951, 1963). Weather maps also are valuable when correlating a season's data with general weather conditions. Such correlations may lead to a better understanding of hawk migrations.

Several sources of weather maps are available. Television stations and newspapers prepare reasonably good maps which contain sufficient detail on the positions of cold fronts and high- and low-pressure areas to make fairly accurate predictions of hawk flights. Weekly booklets of weather maps also are available on a subscription basis from the federal government. These are now issued at the end of each week rather than daily.

DECOYS

The value of decoys in luring hawks to a shooting stand was well known to the gunners who frequented Bake Oven Knob and Hawk Mountain during the 1930's. The most effective decoys were live pigeons placed in cloth harnesses and suspended from long poles with a wire or string. Sharp-shinned Hawks were especially attracted—particularly when the pigeon was lifted from its perch and flapped its wings.

At Bake Oven Knob, a large papier-mâché Great Horned Owl is placed on a long pole in an upright position. The size and coloration of this decoy is unimportant since Sharp-shinned Hawks are readily attracted to decoys of various sizes and colors. Other species occasionally can be decoyed closer to the lookout, too.

Decoys are an aid to photography and also help to assure more complete hawk counts by bringing into view some birds which otherwise might be unseen.

RADIO COMMUNICATION

For years hawk watchers considered the value of radio communications between lookouts. In 1967, Hawk Mountain Sanctuary used radios to study migration routes of hawks crossing a small section of eastern Pennsylvania. A 25-watt transmitter was established in the Sanctuary headquarters building, and Motorola radiophones were used at nearby lookouts. The radios were operated on a For-

Goshawks and other hawks sometimes dart at decoys. The decoys help to bring into view some hawks which otherwise might pass unseen. Photo by Fred Tilly.

A Sharp-shinned Hawk darting past an artificial owl decoy. Photo by Fred Tilly.

estry Conservation Band frequency. One 10-watt unit was positioned on the main Hawk Mountain lookout. The second was operated at Bake Oven Knob, sixteen miles northeast of Hawk Mountain. Reception between the stations was improved by putting a small antenna in a tree. This communications network enabled observers at the upridge site (Bake Oven Knob) to notify observers at Hawk Mountain that eagles or other hawks had been seen there and might be approaching Hawk Mountain. When easterly and southerly winds occurred, evidence suggested that many birds seen at Bake Oven Knob failed to reach Hawk Mountain. Some data also were gathered on hawk flight speeds along this section of the Kittatinny Ridge. Unmarked birds limited the accuracy and validity of the project's data, however.

Radio communication between lookouts was a novel experiment but failed to provide much not already available from hourly records of the two sites. Radios are more useful to hawk watchers using several lookouts within Hawk Mountain Sanctuary by permitting rapid coordination of field studies. Haugh (1972) also used walkie-talkies in his hawk migration studies.

ELECTRONIC OBSERVATIONS

Within the past decade much interest has developed in the use of biotelemetry techniques for studying wildlife movements. Most of this work is restricted to game animals. Dunstan (1972) summarized biotelemetry techniques for raptor studies.

Radar also is used in ornithological research with increasing regularity. However, there are only two or three locations in North America where radar has been used to study hawk migrations, and data from these sites either are unpublished or are inconclusive. The best summary of the use of radar in ornithological research is by Eric Eastwood (1967) in *Radar Ornithology*.

OBSERVATIONS FROM AIRCRAFT

The potential value of aircraft as a base for studying hawk migrations is relatively little exploited. Nonetheless, aircraft might provide observers with many insights into the routes which migrat-

Sailplanes may be useful in studying migrating hawks because they may permit observers to follow the birds as they cross restricted geographic areas. Photo by Donald S. Heintzelman.

ing hawks follow, the altitudes at which they fly, and other related phenomena. Nagy (1973: 23), for example, suggested that sailplanes might have value in studying hawk migrations. Indeed, in East Africa, Pennycuick (1973) already has demonstrated that powered gliders such as the Schleicher ASK-14 are extremely useful in studying the soaring flight of vultures in Serengeti National Park in northern Tanzania. Raspet (1960) also used a sailplane to study the flight of Black Vultures in the southern United States.

The most unusual use of aircraft in studying hawk migrations was suggested by E. I. Stearns (1948a; 1948b; 1949: 110) and conducted in New Jersey with the assistance of James L. Edwards and Alfred E. Eynon. A blimp was used and the technique was called "blawking."

The Urner Ornithological Club of Newark, N.J., selected 21 September 1948 to conduct the blawking experiment. The Tide Water Associated Oil Company supplied the blimp, and a second blimp also was sent aloft later in the day through the cooperation of the Tide Water and associated companies. The day was clear with an air temperature of 65 degrees Farenheit and a 10-mile-per-

The blimp used by the Urner Ornithological Club on 21 September 1948 in its New Jersey "blawking" project. Photo by Edwin I. Stearns.

The gondola of the blimp used in the 1948 blawking project. Photo by Edwin I. Stearns.

hour northwest wind. In addition to observers in the blimps, observers also were stationed on the ground at Upper Montclair, New Jersey, and elsewhere. Ground-to-blimp radio communications were not available, but visual signals were given from the ground stations via fluorescent panels using the following color codes (Stearns, 1948b):

Signal	*Significance*
Red strip	Poor flight
Red, white strip	Fair flight
White, red, white strip	Good flight
Cross	Cross ridge flight
Tee	Ridge flight
Red cross (or Tee)	Hawks mainly Broad-wings
Red, white cross (or Tee)	Mixed species
White cross (or Tee)	No Broad-wings
Yellow strip	Hawks visible in air, high
Yellow-white strip	Hawks visible in air, low
White, yellow, white strip	No hawks visible now

The log of the first blimp follows (Stearns, 1948b):

EDST

12:34 P.M. Took off from Lakehurst.

1:58 Arrived over Montclair, elevation 1700 ft. Read the panels and started NE toward the George Washington Bridge over the Hudson River.

2:18 Sighted 3 Broad-wings; altimeter read 2300, hawks probably at 2800 ft.

2:25 One Broad-wing was seen gliding in the direction of the airship; it was lined up with the window bar and clocked at 32 m.p.h. The airship then circled back over the Quarry and went SW down the ridge.

2:50 A kettle of over 180 Broad-wings was sighted over Crystal Quarry, West Orange (Pal's Cabin), with the center of the flock at an altitude of 2700 ft. Unfortunately the ship was heading away from the direction in which the hawks were drifting, and when they started to peel off and go into a glide they were lost as the ship tried to get turned around. The ship cruised back up the ridge almost to the Quarry.

3:31   Spotted a 50-bird kettle at 2000 ft. just SW of the Quarry and followed them as they peeled off. The average speed of hawks in the glide was 26 m.p.h. The hawks gradually dropped lower and dispersed, most of them going into the woods just north of N.J. Highway 10 and just SW of Vincent's Pond. The distance from the thermal to the roosts (later checked on a map) was 4 miles. The ship cruised around again, circled over the Quarry, and although the panels still read "good flight" with hawks visible, they could not be found from the ship, and it headed for the Ramapo Mountains to attempt to spot a ridge flight there.

3:46   Over Maywood.

4:00   Arrived at Ramapos, found no hawks; left at 4:45, returned over Montclair, Verona, Plainfield, Bound Brook.

6:37   Landed at Lakehurst.

The log of the second blimp (not originally part of the project and sent aloft during the afternoon unknown to the other observers) was as follows (Stearns, 1948b):

EDST

2:45 P.M. Took off from Lakehurst.

4:00   Arrived at Montclair; saw 6 Broad-wings flying south along the First Ridge south of Bloomfield Avenue about 0.1 miles apart. The birds were low and in a ridge flight.

4:05   Flew into a spiraling flock unexpectedly, not having seen them, and dispersed them in all directions. Some birds within 200 ft. of the ship appeared frightened. Counted 24 birds but there may well have been many more. Altitude 1500 ft., position slightly east of Quarry.

4:07   55 Broad-wings spotted over Cedar Grove Reservoir, a considerable distance from the ship. They were easily counted with binoculars, but as the ship continued straight ahead the birds were lost from view and could not be found again. The airship altitude was 1750 ft. and the birds were well above this. The upper limit was estimated as twice the blimp's altitude from the ground, or 2900 ft. (The birds may have been closer than estimated, but certainly no lower than 2400 ft.)

4:10   4 Broad-wings seen separately in southerly ridge flight north of Cedar Grove Reservoir.

4:45   Left Montclair, returned by way of Allentown, Pa.
10:30   Landed at Lakehurst.

Ground observers at Upper Montclair counted 2,150 Broad-winged Hawks, observers in the blimps only 290 hawks. This suggests that hawks are more difficult to locate from the air than from the ground.

Despite the difficulty of spotting hawks from the air, four separate kettles were found spiraling upward at 1,500, 2,000, 2,700, and 2,400–2,900 feet above sea level. The birds were flying above a valley with an elevation of 190 feet and rising over a ridge with an elevation of 590 feet. "It is not known that these were the highest kettles of the day, nor measured at their highest point except for the 2000 foot kettle. Although earlier ground estimates had placed the kettles at greater heights, the maximum height reached may well be only 3000 feet, and many times the birds abandon their upward spiraling and 'peel off,' or enter their straight, downward glide, at only 2000 feet.

"The birds peeling off from the 2000 foot thermal, a rising column * of air heated from a warmer ground area, were successfully followed until they roosted in trees 4 miles away at a ground elevation of about 450 feet. Thus the ratio of glide to fall was about 12 to 1. The air speed of the Broad-wings in the glide was 32 mph in one measurement and 26 mph in another. Judged by the criterion that the hawks were not frightened if they continued their glide in an undeviating line, the birds did not seem to mind the airship provided it was more than 300 feet distant" (Stearns, 1949).

Ground observers overestimated the height which hawks reached in thermals and were unable to judge airspeeds and distance of glides accurately.

TRAPPING AND BANDING MIGRATING HAWKS

Most hawk banding in North America has been confined to nestlings. Several excellent summaries of recoveries † of these birds

---

* See footnote on page 5 and chapter 14.
† The recapture (usually dead) of a banded bird at a site other than where it was banded originally.

A female Sparrow Hawk caught in a dho-ghaza net at a raptor trapping and banding station. Photo by Donald S. Heintzelman.

have been published (Cross, 1927; Lincoln, 1936; Worth, 1936). Little effort has been made to trap and band migrating hawks until recently. An exception is the trapping of migrating Peregrine Falcons for falconry purposes on Assateague Island, Maryland, since 1938 (Berry, 1971). It is doubtful if many of these birds were immediately released, however. Several hawk trapping and banding stations of major importance in North American now operate for long periods of time during autumn, however. Numerous weekend stations also are being devoted to trapping and banding migrating hawks in the East during autumn.

A variety of devices are used to lure and capture migrating hawks. These include bal-chatri traps (Berger and Mueller, 1959), bow-net traps (Meng, 1963; Austing, 1964) and large-mesh mist nets and dho-ghaza traps (Clark, 1971). Occasionally other devices are used. Descriptions of the most important trapping stations follow.

Tri-County Corners, Pennsylvania

This station is operated by C. J. Robertson on the crest of the Kittatinny Ridge in an old field at the junction of the boundaries of Lehigh, Berks, and Schuylkill counties, Pennsylvania. Robertson uses a variety of equipment typical of the devices found at most hawk banding stations. Table 1 in Appendix 2 presents the seasonal totals of hawks trapped and banded here.

Lehigh Furnace Gap-Lehigh Gap, Pennsylvania

These two stations are on the crest of the Kittatinny Ridge in Lehigh County. Lehigh Furnace Gap is an old hawk shooting site used when northerly and westerly winds prevail. A ledge about two miles west of Lehigh Gap is used when southerly or easterly winds prevail. Both stations are operated by John B. Holt, Jr. Table 2 summarizes the combined banding totals for both stations.

Cape May Point, New Jersey

Witmer Stone (1937) documented in depth the importance of Cape May Point as an autumn concentration point for birds. The spectacular autumn hawk flights at the Point received the special attention of Allen and Peterson (1936), Choate (*in* Heintzelman, 1970e), Choate and Tilly (1973), and Stone (1922, 1937). Since 1967, a hawk trapping and banding station has been operated near the lighthouse along the edge of a field and wooded area (W. S. Clark, 1968, 1969, 1970, 1971, 1972, 1973). Yearly banding totals are shown in table 3. The 1971 totals reflect two trapping stations being used whereas one was used previously.

Cedar Grove, Wisconsin

The first hawk trapping and banding station in the United States was established along the west shore of Lake Michigan at Cedar Grove about forty-five miles north of Milwaukee, Wisconsin. In 1921, ornithologists from the Milwaukee Public Museum discovered hawk migrations in the vicinity of Cedar Grove. In 1936 the trapping station was established (Jung, 1964). In 1950, Helmut C. Mueller and Daniel D. Berger began operating a new banding program at the Cedar Grove Ornithological Station (Mueller and Berger, 1961). Additional papers have detailed aspects of Sharp-shinned Hawks observed and banded at Cedar Grove (Mueller and Berger, 1967b), the roles of wind drift * and leading lines † during migration (Mueller and Berger, 1967a), sex ratios and measurements of migrant Goshawks (Mueller and Berger, 1968), and the daily rhythm of hawk migrations (Mueller and Berger, 1973).

Duluth, Minnesota

In 1972 a hawk trapping and banding station was established at the Hawk Ridge Nature Reserve in Duluth, Minnesota, through the efforts of David Evans (Sundquist, 1973: 5). The station began operating on 7 September and continued daily trapping through mid-November. A total of 687 raptors were banded: 392 Goshawks, 160 Sharp-shinned Hawks, 4 Cooper's Hawks, 73 Red-tailed Hawks, 4 Broad-winged Hawks, 1 Golden Eagle, 9 Marsh Hawks, 2 Pigeon Hawks, 15 Sparrow Hawks, 12 Great Horned Owls, 10 Long-eared Owls, and 5 Saw-whet Owls. The number of Goshawks handled probably represents the largest living sample of that species ever examined critically during a single migration season anywhere.

Hawk Cliff, Ontario

Hawk Cliff, located on the northern shore of Lake Erie in Elgin County, Ontario, has been known as a hawk flyway since about 1931 (Field, 1970; Haugh, 1972; Sutton, 1956), but it was not used as a trapping and banding station until the autumn of 1969. It is now used regularly in autumn to trap and band migrating Sharp-shinned Hawks, Sparrow Hawks, and various other species (Field, 1970, 1971).

* "The displacement of a bird due to wind" (Mueller and Berger, 1967a).
† "Topographical features, usually long and narrow, with characteristics that induce migrating birds to follow them. The birds are influenced by these lines in choosing their direction of flight, being so to speak led by them" (Geyr, 1963).

Point Pelee, Ontario

Since 1955, banders have trapped and banded migrating Sharp-shinned Hawks and other species at Point Pelee National Park, Ontario (Gray, 1961). Gray documents a large flight of Sharp-shins at the Point on 17 September 1960 of which 192 were caught and banded. The following day another large flight developed of which 56 were banded and more than 300 were observed within a few hours. About 15 percent of the hawks banded at the Point are recovered.

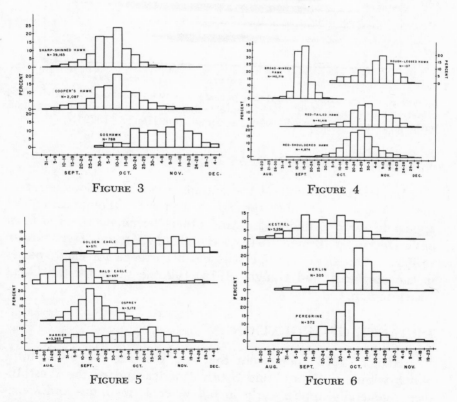

FIGURES 3–6 The temporal migration periods of hawks observed at Hawk Mountain, Pa., from 1954 through 1968. Similar periods apply to hawks migrating past Bake Oven Knob, Pa. Reprinted from Haugh (1972).

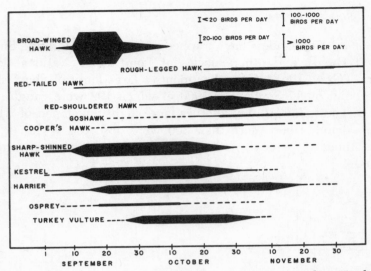

FIGURE 7 The temporal migration periods of hawks observed at Hawk Cliff, Ontario, based upon 1967 data supplemented by additional incomplete data gathered from 1950 through 1966. Reprinted from Haugh (1972).

AUTUMN MIGRATION SEASON

In general, the bulk of the autumn hawk migrations occur in the East from September through November, although stragglers appear in early to mid-August and others terminate the season in early December. Broun (1949a), Haugh (1972), and Heintzelman (1972d) charted the specific seasonal occurrence for each species in the northeast, and Haugh (1972) did the same for the Great Lakes region.

DISTANCE OF MIGRATION

Haugh (1972) points out that long-distance migrants (birds which winter in Central and South America and/or the West Indies), generally appear early in fall whereas medium- and short-distance migrants move later in autumn. Immature Sharp-shinned Hawks migrate earlier than adults.

Haugh (1972: 4) grouped Broad-winged Hawks and Ospreys as long-distance migrants. Turkey Vultures, Sharp-shinned Hawks,

Marsh Hawks are medium-distance migrants. This bird is an adult male. Photo by Fred Tilly.

Cooper's Hawks, Red-shouldered Hawks, Rough-legged Hawks, Bald Eagles, and Marsh Hawks are medium-distance migrants. Goshawks, Red-tailed Hawks, and Golden Eagles are short- to medium-distance migrants. The Sparrow Hawk is a short- to long-distance migrant, and Pigeon Hawks and Peregrines are medium- to long-distance migrants.

2

Hawk Identification

Most people, limited in their experience with birds of prey, find these birds relatively difficult to identify. Hence the large numbers of diurnal * raptors which concentrate at various coastal and inland locations in eastern North America during autumn offer exceptional observational and field study opportunities. Basic information required for correct hawk identification is provided in field guides to bird identification by Peterson (1947), Pough (1951), and Robbins (1966). Supplementing these are two newer guides detailing more subtle points of hawk identification based upon the appearance of the birds during migration (Heintzelman, 1972d; Brett and Nagy, 1973).

This chapter summarizes details of diurnal raptor identification based upon information contained in each of the previously mentioned publications, as well as other literature, and my own extensive experience studying these birds. When considering the size of raptors, remember that they exhibit marked sexual dimorphism. Females are about one-third larger than males. Moreover, the shape of a bird is as important as its pattern of color. Hence, under poor light conditions, many raptors are identified largely by shape and

* Birds of prey (order *Falconiformes*) which are active during the day.

pattern of behavior. Equally important, one should realize that it is not always possible to identify every hawk which is observed.

VULTURES

Two species of vultures occur in eastern North America, the Turkey Vulture and the Black Vulture. Turkey Vultures commonly are observed at most eastern hawk lookouts. Black Vultures rarely are seen except in the southern states.

TURKEY VULTURE

Turkey Vultures are large, nearly eagle-size and eagle-like soaring birds. Observed in flight, they appear black with the underside of their wings grayish or silvery, giving the birds a two-toned appearance. They characteristically hold their wings in a slight V or dihedral, thus aiding identification. On very rare occasions they hold their wings flat and eagle-like which, if seen at a great distance, may cause the birds to resemble eagles. Not infrequently these vultures also hold their naked heads, crimson-red as adults and grayish-black as immatures and sub-adults, downward in contrast to eagles, which hold their heads forward. The Turkey Vulture's tail also extends far beyond the rear edge of the wings. They typically rock or tilt from side to side while gliding or soaring on updrafts * or circling overhead. Their occasional wingbeats are powerful and labored.

BLACK VULTURE

Black Vultures also are large black birds with wingspreads from 54 to 60 inches and whitish patches on the underside of their wings near the tips. Their tail is short and the tip square, barely extending beyond the back edge of their wings when in flight. Their flight style, a particularly good field mark, consists of several rapid flaps followed by a short sail. These vultures rarely are seen at the northeastern hawk lookouts. Nonetheless, they may appear at some southern locations.

* An upward deflection of surface winds caused by striking the sides of a mountain or other topographic feature. Some authors refer to updrafts as deflective air currents.

A Turkey Vulture, approaching head-on, showing the characteristic dihedral in the wings and the typical position of the head. Photo by Fred Tilly.

The overhead flight silhouette of a Turkey Vulture soaring on mountain updrafts. Photo by Donald S. Heintzelman.

ACCIPITERS

These are forest-loving bird-eating hawks. In general, they are characterized by short rounded wings and a long, rudderlike tail. Three accipitrine species are observed as autumn migrants in eastern North America: the Goshawk, the Sharp-shinned Hawk, and the Cooper's Hawk. Only the Sharp-shin is a common migrant. Cooper's Hawks occur in lesser numbers—a proportion of about 25 Sharp-shins to 1 Cooper's according to Brett and Nagy (1973: 45)— and Goshawks rarely are observed in notable numbers except during years of invasions.

The typical accipitrine flight pattern consists of three parts— several rapid wingbeats, a brief period of soaring or sailing flight, then more rapid wingbeats. Experienced observers frequently can detect minor variations in the flight patterns of each species.

GOSHAWK

Goshawks are the largest and most powerful of our accipiters. Adults have wingspreads from 40 to 47 inches and are bluish-gray above and pale grayish-white below. Their dark cap and cheek markings, and white eyebrow stripe, are distinctive field marks. Often, in adult birds, the dark cheek mark is more readily noticed than the white eyebrow stripe. Immatures are brown above and heavily streaked below. They, too, have a conspicuous white eyebrow stripe, their most important field mark, but they lack the dark cheek mark of adults. Goshawks also have long tails, the tip of which is slightly to fairly markedly rounded.

These hawks also have a slower wingbeat than the smaller accipiters, but this characteristic is useful only to experienced observers. However, the unusually conspicuous white undertail coverts * of adult and immature Goshawks also aid identification.

SHARP-SHINNED HAWK

Sharp-shinned Hawks are the smallest, and most common, of our accipitrine species. They have wingspreads from 20 to 27 inches,

* Small feathers which cover the bases of wing- and tail-flight feathers.

Upper left, a lateral view of an adult Goshawk showing the dark cap and cheek patch, and the white eyebrow stripe. Photo by Fred Tilly. *Upper right,* an adult Goshawk showing the conspicuous white undertail coverts and typical accipitrine shape. Photo by Fred Tilly. *Lower left,* the overhead flight silhouette of an adult Goshawk. Photo by Donald S. Heintzelman. *Lower right,* an immature Goshawk showing the white eyebrow stripe. Photo by Donald S. Heintzelman.

and the tip of the tail of a Sharp-shin can be square, slightly notched, or slightly rounded (Forbush, 1927; Roberts, 1932). All three variations are seen. Hence a large female Sharp-shin, with a slightly rounded tail, can be confused with a small male Cooper's Hawk. However, the Sharp-shin's wingbeat is fairly rapid. Observers should become familiar with the shape and wingbeat of this species and use them as bases for comparison with hawks of questionable

Upper left, a Sharp-shinned Hawk (immature) showing the notch in the tip of its tail and overhead flight silhouette. Photo by Fred Tilly. *Upper right,* a Sharp-shinned Hawk (adult) showing another variation in tail shape. The tip of this bird's tail is square. Photo by Fred Tilly. *Bottom,* another variation in Sharp-shinned Hawk tail shape. The tip of this bird's tail is slightly rounded; it is a large immature female—not a Cooper's Hawk, with which it might be confused. Photo by Donald S. Heintzelman.

identity. Brett and Nagy (1973: 44) also state that Sharp-shinned Hawks appear to have shorter heads, which are tucked into their shoulders, in comparison with the longer-headed Cooper's Hawk. I have not found this to be a particularly helpful field mark, and much remains to be learned about field identification of large Sharp-shinned and small Cooper's Hawks. These two species usually cause the most difficulty in respect to correct identification.

COOPER'S HAWK

Cooper's Hawks are intermediate in size among our three accipitrine species. Their wingspreads range from 27 to 36 inches.

Left, an adult female Cooper's Hawk showing the extremely rounded tip of its tail which is characteristic of this species. The wings are extended fully to achieve maximum lift. Photo by Donald S. Heintzelman. *Right,* an adult female Cooper's Hawk soaring on updrafts. The primaries are not extended fully. Photo by Donald S. Heintzelman.

The *extremely rounded* tip of the tail is the most important field identification mark. In addition, these birds have a slightly slower wingbeat than the more common and smaller Sharp-shinned Hawk —the species with which Cooper's Hawks are most likely to be confused. Although Brett and Nagy (1973: 44) suggest that Cooper's Hawks appear longer-headed than Sharp-shins, I am not satisfied that this is a completely acceptable and accurate field mark. Hence, in my opinion, the only reliable field mark, useful in separating questionable Sharp-shins from Cooper's Hawks, is the latter's extremely rounded tip of the tail.

BUTEOS

Buteos are soaring hawks. They may occur either in open country, along coastal areas, or in wooded areas depending upon the species and season under consideration. As autumn migrants, most buteos occur along inland ridges and the Great Lakes shorelines, rather than the Atlantic coastline, due to the more favorable updrafts prevailing at the former locations. Four species regularly occur at various eastern concentration points during autumn: the Red-tailed, Red-shouldered, Broad-winged, and Rough-legged Hawks. The Swainson's Hawk is rarely observed at various eastern hawk lookouts.

Buteos commonly circle high overhead with their wings and tail fully spread to achieve maximum lift. However, during their autumnal migrations each of the species frequently closes the tail feathers, and fold in (rather than extend) the primary flight feathers, while gliding or soaring on updrafts. This gives the birds a more streamlined appearance.

RED-TAILED HAWK

Red-tailed Hawks, with wingspreads from 46 to 58 inches, are the most variable of the four buteos in respect to color. Both albino and completely melanistic individuals occasionally are seen during migration. These individuals are extremely rare, however. More normally colored Red-tails, both adults and immatures, have conspicuous bands or dark markings across the belly although, in some individuals, the belly band is seen to be very faint. On very rare occasions an individual may be seen with no apparent belly band. Most individuals are well marked and obvious, however. Hence the belly band, along with the red tail of adult birds, is the most important field mark for the species.

Another helpful aid to Red-tail identification is a characteristic light area near the wrist on the leading edge of each wing when

Left, a Red-tailed Hawk soaring on mountain updrafts. The wings are not extended fully. This helps to streamline the bird. The prominent belly band is an important identification mark. Photo by Donald S. Heintzelman. *Right,* a lateral view of a migrating Red-tailed Hawk. Many migrating Red-tails appear in this position at hawk lookouts. Photo by Donald S. Heintzelman.

A Red-shouldered Hawk in a typical overhead flight position soaring on mountain updrafts. The large areas of each wing through which light passes are called "windows." They are helpful identification marks but are not restricted entirely to this species. Photo by Donald S. Heintzelman.

the bird is seen head-on. As the hawk approaches, an observer has the impression that the bird has a headlight on each wing. This often permits rapid and accurate identification of an approaching Red-tail—even at considerable distances. Additional aids to identification include the obvious red tail of adults, and their habit of hanging motionless, or sometimes hovering briefly in mid-air, while hunting. One frequently observes apparently local (nonmigratory) birds engaging in this activity in the vicinity of various hawk lookouts. Finally, the cere * also appears light on many Red-tails but not as conspicuous as on Broad-winged Hawks.

RED-SHOULDERED HAWK

Adult Red-shouldered Hawks, with wingspreads from 32½ to 50 inches, are among the most colorful of our eastern raptors and are easily identified when observed at close range and under good light conditions by their banded tail and reddish shoulder patch

* The basal covering of the upper mandible. In hawks it is soft and sometimes brightly colored. The nostrils are within, or at the edge of, the cere in hawks.

on the upper side of each wing near the wrist. Under less favorable conditions these birds are more difficult to identify. Nonetheless the conspicuous "windows" * are useful aids although Red-tails, Broad-wings, and other species occasionally show similar windows when observed under some light conditions. The lack of a belly band, coupled with its fairly long tail, and moderately large size also are helpful in identifying Red-shouldered Hawks. Immature birds are fairly heavily streaked on their undersides. Red-shouldered Hawks frequently exhibit butterfly-like or fluttering wingbeats followed by a period of sailing flight which may suggest, to some degree, an accipiter's flight pattern.

BROAD-WINGED HAWK

The Broad-winged Hawk is the smallest of our buteos. Its wingspreads range from 32 to 39 inches. They are identified readily as adults by their two conspicuous white and two dark tail bands. Immature birds resemble other immature buteos but can be separated from them by their smaller size. Broad-wings also have fairly conspicuous black wingtips and are proportionally more chunky than other buteos. Seen head-on, a large light (pale yellow at close range) area behind the bird's bill (the cere) creates a single headlight effect. This is noticeable both on adult and immature birds (Red-tails also show this characteristic to a lesser degree). Broad-wings also hold their wings in a slightly bowed position when gliding or using updrafts for soaring. In the East, these are the only hawks which flock in large numbers in kettles—an excellent aid to field identification in late August, September, and very early October. One should be aware that other hawks occasionally use thermals, sometimes along with groups of Broad-wings, however. Hence one should remain alert for other species which may be mixed among kettling Broad-winged Hawks.

ROUGH-LEGGED HAWK

Rough-legged Hawks are rare autumn migrants at most eastern concentration points, and usually appear late in the season. Their wingspreads are from 48 to 56 inches. They are extremely variable

* Areas of translucence in the wings near the wrist.

Top, a kettle of Broad-winged Hawks inside a thermal. Photo by Alan Brady. *Lower left,* an adult Broad-winged Hawk circling overhead. Its wings and tail are spread fully to achieve maximum lift. The black and white tail bands are important identification marks; so, too, are the bird's chunky proportions. Photo by Donald S. Heintzelman. *Lower right,* an adult Broad-winged Hawk gliding from a thermal. Photo by Donald S. Heintzelman.

A light-phase Rough-legged Hawk. Photo by Karl H. Maslowski.

in color. Cade (1955), for example, has shown that North American Rough-legs exhibit a continuous range of color variation between so-called light and dark phases. For field identification purposes, light- and dark-color phases can be recognized, as is done in the standard guides to bird identification, however. Individuals in both phases have a broad dark terminal or subterminal band on a white tail. Cade (1955) further has shown that variations in the tail pattern in adult Rough-legs reflect sexual dimorphism in many examples. Light-phase birds also have a black belly band and black wrists on each wing. Dark-phase birds are largely dark but show extensive white areas on the underside of each wing. Rough-legs are unusual among buteos in that they hover regularly, although one does not normally see them doing so during migration.

SWAINSON'S HAWK

Swainson's Hawks are rare at most eastern hawk lookouts although a few sightings have been reported during recent years. They occur both in a light- and a dark-color phase. Dark birds are rare but are almost completely black although the undersides of the wings appear dusty or mottled. Light-phase birds are more typical and are distinctive due to the dark head, back, primaries, and chest, which contrast sharply with the light belly. Individuals in both color phases have a dark, conspicuous terminal band on the tail.

With wingspreads ranging from 47 to 57 inches, in flight Swainson's Hawks hold their wings in a slight V or dihedral similar to that characteristic of Turkey Vultures and Marsh Hawks. Any bird observed in the East and suspected of being a Swainson's Hawk should be examined critically.

EAGLES

Two species of eagles occur in the East—the Golden Eagle and the Bald Eagle. Both are very large, majestic birds.

GOLDEN EAGLE

Adult Golden Eagles, with wingspreads from 75 to 94 inches, show little or no white in their tail and essentially are dark brown birds (black when seen at a distance). Normally the golden hackles are not visible. When the birds are seen at close range, and under good light conditions, these golden nape feathers clearly are obvious—a splendid sight. Birds from the Appalachian population, the eagles observed at eastern hawk lookouts, sometimes have whitish markings similar to the headlights on Red-tailed Hawks, on the leading edge of their wings.

Juvenile, immature, and sub-adult Golden Eagles are more vividly colored than adults. They have a conspicuous white area on the upper and lower surfaces of their wings and a broad, white basal area on the tail contrasting with a wide dark terminal band. The sub-adult plumage of the Golden Eagle can resemble closely the Bald Eagle's immature plumage. But sub-adult Golden Eagles have little white on the basal portion of their tail, although some white is retained.

Finally, Golden Eagles of all ages and sexes have proportionally smaller heads and bills than Bald Eagles. Experienced observers sometimes find this feature a useful aid to identification.

An adult Golden Eagle (photographed in New Mexico) showing a typical overhead flight silhouette. Photo by Walter R. Spofford.

BALD EAGLE

Adult Bald Eagles, with wingspreads from 72 to 98 inches, are unmistakable due to their conspicuous white head and tail, which contrast vividly with their dark brown body (black when seen at a distance). Immature birds are more difficult to identify since they frequently are confused with sub-adult or adult Golden Eagles. Although immature Bald Eagles are variable in color, they usually

A juvenile Golden Eagle (photographed in New Mexico). The conspicuous white patches in the wings, and the broad white band on the tail, are important field marks. Photo by Walter R. Spofford.

have a white or mottled underwing lining extending from the body outward toward the tip of the wing. The white basal area on the tail also is much less extensive than that on the immature Golden Eagle's tail. Bald Eagles have much larger heads and bills than Golden Eagles, too.

Left, an adult Bald Eagle soaring on mountain updrafts. The dark body and white head and tail are distinctive. Photo by Donald S. Heintzelman. *Right,* an immature Bald Eagle showing the whitish underwing linings and faint band on the tail. Photo by Donald S. Heintzelman.

For the purpose of determining the age of migrant Bald Eagles, which cannot be handled and examined critically, it seems best to separate the birds into two general categories—adults and immatures. Adult birds thus would be all individuals with white heads and tails; all other individuals would be lumped together as immatures. In point of fact, however, Brown and Amadon (1968) and Southern (1964) have shown that the full adult plumage of the Bald Eagle may not be acquired until the birds are six or more years old.

HARRIERS AND OSPREYS

MARSH HAWK

The Marsh Hawk, whose wingspread is from 40 to 54 inches, is the only harrier * occurring in North America. It has a relatively long tail and fairly long, narrow wings which frequently are held in a slight vulture-like dihedral. Although the wings sometimes appear slightly rounded at the tips, during migration they frequently appear slightly pointed (almost falcon-like). Individuals of all ages dis-

* A term used more commonly in the Old World for hawks in the genus *Circus.*

Upper left, the slight dihedral in the wings, and the white rump patch, are important identification marks for the Marsh Hawk. Photo by Fred Tilly. *Upper right,* an adult male Marsh Hawk showing the pattern of its plumage. Photo by Harry Goldman. *Lower left,* an adult female Marsh Hawk showing the streaking on its breast. Photo by Harry Goldman. *Lower right,* an immature Marsh Hawk soaring on mountain updrafts. The underparts of its body are unstreaked. Photo by Fred Tilly.

play a conspicuous white rump patch, which is the best identification mark for the species. The Marsh Hawk's flight style also is characteristic, consisting of an unsteady rocking, tipping, or zigzag pattern.

Adult male Marsh Hawks are gray with black wingtips whereas adult females are brown and heavily streaked on their ventral or lower surface. Immatures of both sexes are brown with cinnamon bodies and no streaking.

OSPREY

Ospreys are among the most majestic of the diurnal raptors forming part of the autumn migrations. They are large birds (wingspreads 54 to 72 inches), dark above and whitish below, with a black wrist mark on the underside of each wing. During migration they frequently bend their wings into a conspicuously crooked position when soaring on updrafts—an extremely characteristic field mark. Seen head-on, Ospreys also typically hold their long wings in an arched position which produces a noticeable bow in the flight profile of the bird as the wings bend downward. This is extremely useful in permitting identification of Ospreys at great distances. On infrequent occasions, these birds can be confused with adult Bald Eagles if observed at great distances and under poor light conditions. However, the general shape and white body of the bird usually

A migrating Osprey with its wings folded into a crooked position. This overhead flight silhouette is extremely characteristic of migrating Ospreys. It is an excellent identification mark. Photo by Donald S. Heintzelman.

enable correct identification after careful observation. Eagles have dark rather than white bodies.

FALCONS

Falcons have long, pointed wings and relatively long tails—a combination which gives them a characteristic silhouette. They generally fly with strong, rapid wingbeats although considerable variations occur in the flight style of the various species.

GYRFALCON

These splendid Arctic falcons are *extremely rare* at most eastern hawk lookouts. At Hawk Mountain, for example, Broun (1949a: 167–168) recorded Gyrfalcons only six times, three of which were hypothetical. They are our largest falcons (wingspread 44 to 52 inches), occurring in white-, gray-, and black-color phases. Gray birds are most frequently observed in the East. A Gyrfalcon's wingbeat is relatively slow and gull-like although their flight is fast. However, Christensen et al. (1973: 104) state that their flight is slower than a Peregrine's, the wingbeats being ". . . slower and

An adult Peregrine Falcon showing the characteristic pointed wings. The conspicuous cheek mark also is visible. Photo by Karl H. Maslowski.

shallower, appearing almost as if made by hands alone." When soaring and gliding, the wings extend straight from the body and are held level or slightly bowed although occasionally a slight dihedral profile is assumed, suggesting that of a harrier.

PEREGRINE FALCON

This splendid falcon (wingspread 38 to 46 inches) is identified best by its long, pointed wings and dark facial pattern, dark cap, and its quick "rowing" flight style. Adults have slate-colored backs but are pale below. Immatures are brown and streaked heavily below. Extreme caution should be exercised in identifying a falcon as a Peregrine since all North American Peregrine populations are endangered.

PIGEON HAWK

Pigeon Hawks are small, dark, roughly jay-size birds with wing-spreads from 23½ to 26½ inches. They suggest somewhat minia-ture Peregrine Falcons. Adult males are bluish-gray above with several prominent black tail bands. Adult females, and immature

Left, a Sparrow Hawk sailing overhead. Notice the long pointed wings, typical of the falcons. Photo by Donald S. Heintzelman. *Right,* a Sparrow Hawk soaring overhead. The wings are folded back to achieve streamlining. Photo by Donald S. Heintzelman.

birds, are brown. The birds are heavily streaked below. Not infrequently Pigeon Hawks fly low over the ground or treetops, and exhibit a rowing-like wingbeat.

SPARROW HAWK

Sparrow Hawks are our smallest (wingspread 20 to 24½ inches) and most colorful falcons. They also are the most abundant of our migrant falcons. Males have blue-gray wings, females brown wings. Both sexes have conspicuous facial markings. This is our only falcon which hovers, although it seldom does so during migration. The flight pattern frequently is more buoyant and erratic than that of the larger falcons. Sometimes it folds back its wings and then assumes a sickle shape (Brett and Nagy, 1973).

Part 2
The Hawk Lookouts

3

Eastern Canada
Hawk Migrations

Eastern Canada is a major breeding ground for many species of hawks. However, little detailed information is available regarding hawk migrations in this part of North America aside from the scattered literature concerning hawk flights along the shorelines of the Great Lakes, which is discussed in chapter 4.

LABRADOR

Todd (1963) mentions a variety of scattered hawk migrations in various sections of Labrador but does not mention any well-defined flight lines. Probably none exist, but this province is so large, and portions so remote, that some concentration points could easily be overlooked. Austin (1932) also does not provide any major information on Labrador hawk migrations.

QUEBEC

Autumn hawk migrations crossing Quebec seem to be scattered, broad-front movements. Each year, however, the Ornithological Club of Quebec reports some data on migrating hawks in *Bulletin*

Ornithologique. Table 4 summarizes typical Quebec hawk counts for the period September through November of 1970.

NEW BRUNSWICK

Knowledge of autumn hawk migrations in New Brunswick still is fragmentary. Squires (1952) specifically states that Sharp-shinned Hawk migrations occur from late August through mid-September in the vicinity of Point Lepreau in the Bay of Fundy region, however. He also records various other hawks, eagles, and vultures (as accidentals *) at scattered New Brunswick locations during autumn, but no major concentration points are mentioned. A surprising number of hawks and eagles have been seen or collected on Grand Manan Island during autumn.

NEWFOUNDLAND

Little is known about hawk migrations in Newfoundland. Peters and Burleigh (1951) mention some migrations, but no specific information is presented aside from an observation of nine Sharp-shinned Hawks flying past Tompkins on 18 September 1946. This is in the extreme southwest of Newfoundland. Austin (1932) provides no major information regarding hawk flights.

NOVA SCOTIA

Autumn hawk migrations have been more carefully studied in Nova Scotia than in any other section of eastern Canada except for the Great Lakes flights. Hawkes (1958), for example, describes his observations on Brier Island on 2–3 October 1958: "On the western end of the island on October 2nd and 3rd, we found instead of shore birds great numbers of hawks. There was one flock of sharp-shinned hawks numbering about 70, and a flock of perhaps 300 buteos, all the same size. I think they were broad-winged hawks. Both of these flocks were wheeling high overhead as if uncertain of the direction

---

* An individual of a species observed unexpectedly far outside of its normal geographic range.

they should follow. They could only cross the Bay of Fundy to the Maine shore or cross the Gulf of Maine to Cape Cod.

"We identified with fair to complete certainty the following species: marsh hawk, sharp-shinned hawk, goshawk, Cooper's hawk, sparrow hawk, pigeon hawk, broad-winged hawk, and red-tailed hawk. At times a half dozen of two or three species were flying around us at once. Brier Island really is a concentration point for birds in their southern migration."

These migrations long have been known to Wickerson Lent (personal communication), the lighthouse keeper on Brier Island. He reports Broad-wings migrating by the thousands across the island during autumn, for at least the past forty years. He saw only a few hundred during the autumn of 1969, however. Broad-wings rarely appear in spring. Tufts (1962: 120–121) has never seen a live Broad-wing in Nova Scotia and records only one breeding record for the island. Commenting on data received from Wickerson Lent, Tufts writes: "The source of these migrations presents something of an enigma, and this fact tends to rule out the likelihood that the migrants are members of our sparse summer population. On the other hand it is not an uncommon summer resident in New Brunswick (Squires, 1952), which suggests the possibility of an entry from there into Nova Scotia across the interprovincial border through Cumberland County. Reference to the map, however, indicates a more normal route southward for these New Brunswick birds would be through Maine. Aside from being more direct, such a course would eliminate the necessity for an overwater flight across the Bay of Fundy. However, migrating hawks seem to enjoy riding the thermal air currents which are ordinarily encountered along high land ridges. This suggests the possibility that New Brunswick Broad-wings long ago, in establishing their fall migration route, were influenced by the proximity of the North Mountain range in Nova Scotia. This begins in Blomidon, in Kings County, only a short distance by air from the interprovincial border and extends unbroken to Brier Island, where it dips into the sea."

The Broad-winged Hawk migrations passing Brier Island require much additional study. (They resemble similar migrations in the Florida Keys, where Broad-wings have been observed flying over the waters of the Straits of Florida and even over the Dry Tortugas.) The fate of these migrants is unknown. After passing Brier Island, do they perish if they do not quickly reach land again? Radar studies here, and in the Florida Keys, might provide information on the movements of Broad-winged Hawks over open waters.

On the other hand, various other species of diurnal raptors at least occasionally use an over-the-water route between Nova Scotia and the Maine coast. For example, William C. Townsend (letter of 1 October 1973) states that, on several occasions, usually during the morning, he observed Marsh Hawks, Ospreys, Pigeon Hawks, and Sparrow Hawks at sea between Mount Desert Island and Nova Scotia. The hawks were flying about 150 feet above the water and were approaching the Maine coast.

4

Great Lakes Hawk Migrations

In contrast to eastern Canada, the shorelines of the Great Lakes also are major autumn hawk migration routes. Flights occur primarily along the northern or western shorelines (Haugh, 1972). A general overview of these flights was published by Pettingill (1962), whereas Haugh (1972) treated certain of the flights in depth.

Enormous September flights of Broad-winged Hawks are particularly notable in the Great Lakes region. Not infrequently these migrations exceed by two or three times the numbers of Broad-wings observed along the Appalachian ridges in Pennsylvania and elsewhere.

ONTARIO (CANADA)

In southern Ontario, in close proximity to the northern shorelines of the Great Lakes, a number of excellent hawk concentration points exist during autumn. Gunn (1954) states that a low-pressure system across northern Ontario and Quebec, coupled with the passage of a cold front crossing southern Ontario, light northwesterly winds, and good thermal activity indicated by cumulus clouds, produces good September Broad-wing flights. A good variety of other migrating hawks also occur at many of the best-known spots.

AMHERSTBURG

This site is located along the northwestern side of Lake Erie on the Detroit River near Detroit. Pettingill (1962: 45) states that many hawks may be seen from a point of land projecting into the Detroit River opposite the south end of Grosse Ile. The spot is reached by driving west from the center of town. According to Millie Reynolds (report of 16 September 1954), the grounds of the Malden School on Highway 18 east of Amherstburg also is satisfactory as a hawk lookout. Table 5 summarizes the 1954 hawk counts from the Malden School grounds compiled from data filed at Patuxent Wildlife Research Center, Laurel, Maryland. Another favorable observation point is east of Amherstburg on Highway 18 just opposite the lighthouse on Boblo Island (Millie Reynolds, letter of 21 September 1954).

COBOURG

Pettingill (1962: 45) states that "heavy" hawk flights are seen from Cobourg westward. The site is located about midway on the northern shore of Lake Ontario.

HAMILTON

According to Pettingill (1962: 45) hawks follow the north shore of Lake Ontario to Hamilton at the western end of the lake. They then fly cross-country to Lake Erie, where they are joined by more hawks approaching from the north.

HAWK CLIFF

Hawk Cliff, along the northern shore of Lake Erie, is one of the best autumn concentration points in the Great Lakes region for migrating hawks (Cameron, 1964; Clendinning, 1954; Haugh, 1972; W. D. Sutton, 1956). It is located about two miles east of Port Stanley at the south end of a road ending at the brink of a 100-foot-high cliff overlooking Lake Erie (Haugh, 1972: 15).

Haugh discovered that variations in habitat divided hawk flights into three separate flight lines. Falcons usually follow the lake shoreline and cliff closely, passing over fields along the edge of the cliff, whereas accipiters occur one-quarter mile inland along the edge of a wooded ravine and buteos and harriers use varied flight lines. Broad-winged Hawks are the most abundantly seen species at Hawk Cliff, although Red-tails, Sharp-shins, and Sparrow Hawks also occur in fairly large numbers. Table 6 summarizes the 1967 autumn hawk count at Hawk Cliff (Haugh, 1972: 16). This was a relatively poor year for Broad-wings, according to Haugh, who counted 31,210 during three days in 1968. Over 49,000 Broad-wings were seen by observers in two days in 1952 at Hawk Cliff, and in 1961 more than 70,000 were counted in one day (Haugh, 1972: 15).

Although northerly winds sometimes result in large hawk flights at Hawk Cliff, west-southwesterly winds also produced some large flights. Haugh (1972) states that passage of a low-pressure area across the eastern Great Lakes almost invariably precedes by a day or two major hawk flights at Hawk Cliff.

PORT CREDIT

Pettingill (1962: 45) suggests that hawk watchers can use the Route 10 bridge over the Queen's Way. To reach Port Credit from Queen Elizabeth Way, fifteen miles southwest of Toronto, turn off onto Provincial Route 10 and continue to the bridge. Table 7 summarizes Port Credit hawk counts based on data gathered by Lucy McDougall and filed at Patuxent Wildlife Research Center.

POINT PELEE NATIONAL PARK

Point Pelee National Park is located in southern Ontario near the western end of Lake Erie. As mentioned in chapter 1, it is an excellent spot to witness migrating Sharp-shinned Hawks, and many have been trapped and banded here since 1955 (Gray, 1961). Many other diurnal birds of prey also form part of the autumn hawk flights at Point Pelee. Although many raptors avoid crossing large bodies of water, Kleiman (1966) observed a few Rough-legged Hawks flying south off the point across Lake Erie on 6 December 1964. Additional comments on this and related phenomena are presented in chapter 16.

PORT STANLEY

See Hawk Cliff.

SARINA

Pettingill (1962: 45) states that hawks frequently are observed passing over and below the Blue Water Bridge linking Sarina, Ontario, with Port Huron, Michigan. These border cities are located below the southern tip of Lake Huron.

TORONTO

Records filed in 1957 at Patuxent Wildlife Research Center by W. W. H. Gunn demonstrate that moderately good hawk flights were seen from the top of the Imperial Oil Building in Toronto. Table 8 summarizes these counts.

MICHIGAN

Although Cook (1893) and Barrows (1912) made a few comments about hawk migrations in Michigan, much remains to be learned about hawk flights in this state. Pettingill (1962: 45), for example, states that "The majority of hawks in eastern Ontario (that is, the section of the province between James Bay and Lake Huron) presumably take a course south and westward that leads across the narrows of the St. Mary's River near Sault Ste. Marie and into Upper Michigan, thence overland to the shore of Lake Michigan. This they follow westward and southward into Illinois, meanwhile keeping on the west side of Green Bay and cutting across the base of the Door Peninsula.

"There is no evidence that hawks, on reaching Upper Michigan, commonly go into Lower Michigan across the Straits of Mackinac. Strange as it may seem, this stretch of water, whose minimum width is slightly less than five miles, is apparently of sufficient extent to discourage regular passage, except possibly by a few accipiters, eagles and falcons.

"Because few people have watched hawk migrations in eastern Ontario and Upper Michigan, I know of no observation spots that can be pinpointed but from accounts scattered in ornithological literature, there is a strong indication that along the north shore of Lake Michigan between St. Ignace and Escanaba there are places where mass movements can be seen from U.S. Route 2 as it parallels the lake shore." Later, Sheldon (1965) demonstrated that autumn (and spring) hawk migrations cross the Straits of Mackinac.

Wood (1951) also states that moderate to good autumn hawk migrations occur in Michigan's Upper Peninsula, in particular at Isle Royale. Moreover, in the Lower Peninsula, Wood (1951) states that large hawk flights have been recorded at Muskegan.

DETROIT REGION

Organized interest in autumn hawk migrations in the Detroit region began in September of 1951 in response to a request from the U.S. Fish and Wildlife Service to the Detroit Audubon Society to supervise a count of Broad-winged Hawks and other raptors (Miller, 1952). Observations were made between 15 September and 6 October. The largest Broad-wing flights occurred on 26, 27, and 28 September. Miller (1952: 79) states that "In all, 21 persons participated in the count, reporting from 26 locations. A total of 10,035 individual hawks was reported, of which 6,849 represented 14 species, while 3,186 were unidentified. By far the largest number identified were Broad-winged Hawks (6,431 individuals), with Sharp-shinned Hawks (198 individuals) the next most numerous." In 1953 and 1954, Merriam (1953, 1954a, 1954b, 1956) summarized in detail hawk migration studies in the Detroit region. Many Broad-wings crossed Grosse Ile in the Detroit River south of Detroit.

During 1954, for example, Merriam (1954, 1956) reported that the Grosse Ile crossing spread over a five-mile-wide path. After crossing the Ile, the birds climbed to great heights and were lost from view. However, "During the eleven days of September 17th thru 27th, when Grosse Ile was under constant watch, a steady flow of low-flying, small accipiters and falcons was observed crossing almost constantly at the rate of about one every six to eight minutes. Differing from the buteos, these birds continued almost due west and were picked up 20 miles west of Grosse Ile at Rawsonville Road, the boundary of our Region. Their frequency was approximately the same as at the island. It is also interesting that the few Ospreys and Bald Eagles sighted on the mainland were also flying due west, oddly enough, away from Lake Erie."

Table 9 summarizes some of the Detroit region hawk counts, which apparently include counts from adjacent Ontario, based upon the published data of Miller (1952) and Merriam (1953, 1954a, 1954b, 1956).

ISLE ROYALE

According to Wood (1951), the Sharp-shinned Hawk is an abundant autumn migrant at Isle Royale from mid-August to the third week in September. More specific data are not provided.

MUSKEGAN

Wood (1951) states that thousands of Sharp-shinned Hawks migrated past Muskegan between 8–27 September 1929 according to Frank Antisdale.

WISCONSIN

Although Cory (1909) made general comments regarding autumn hawk migrations in Wisconsin, not until the autumn of 1921 was the west shore of Lake Michigan known as a major hawk flyway (Jung, 1935). The area of prime importance is a dune area one-quarter to three-quarters of a mile wide extending for about sixteen miles from about three miles north of Port Washington to three miles south of Sheboygan (Jung, 1964). Particular attention has been devoted to studying hawk flights near Bar Creek at the Cedar Grove Ornithological Station (Jung, 1935; Mueller and Berger, 1961). Pettingill (1962: 44) maps the route of hawks appearing here as passing along the northern shore of Lake Michigan thence southward along the lake's west shore through Cedar Grove. Mueller and Berger (1967: 412–414) also comment briefly on hawk migrations in the upper Great Lakes area. The reader is referred to their paper.

In addition to the autumn hawk migrations in Wisconsin, Mueller and Berger (1965) documented an unusual summer movement of between 300 and 1,000 Broad-winged Hawks on 26 June 1960 at Washington Island off the northern tip of the Door County peninsula in northeastern Wisconsin. It was suggested that these were mainly nonbreeding birds although the significance of this unusual movement is unknown.

CEDAR GROVE

On 25 September 1921, Clarence S. Jung and H. L. Stoddard watched a flight of about 2,040 hawks, mainly Sharp-shins and Cooper's along with seven other species, near Bar Creek, Sheboygan County, about forty miles north of Milwaukee. The flight line in this area is narrow, from one-quarter to one-half mile wide (Jung, 1935). This area is now known as the Cedar Grove Ornithological Station. Mueller and Berger (1966) provide a map and description of the site.

Much of the early hawk watching at Cedar Grove was done by Clarence S. Jung, H. L. Stoddard, and other members of the staff of the Milwaukee Public Museum (Jung, 1964). Since 1950, Helmut C. Mueller and Daniel D. Berger have operated a trapping and banding station at Cedar Grove (see chapter 1). A series of reports detailing various aspects of hawk migrations at this site have been published (Mueller and Berger, 1961, 1967a, 1967b, 1968).

In their detailed analysis of weather and hawk migrations at Cedar Grove, Mueller and Berger (1961) determined that about 92 percent of the migrations occurred, after the passage of a cold front across the Great Lakes region, on days with prevailing westerly winds. Such winds tend to concentrate hawks along the west shore of Lake Michigan.

"The fact that correlations of hawk migration with individual factors associated with frontal passage are not so good as correlations with cold fronts per se suggests that these factors act additively in effecting migration. An alternative suggestion is that weather affects migration indirectly in that it acts to modify, not produce, migration. We propose that the relation of fall hawk migration to cold fronts and winds of 15–25 mph is, simply, a correlation with the occurrence of conditions suitable for updraft formation and, hence, with good conditions for soaring and gliding" (Mueller and Berger, 1961: 189). Table 10 summarizes Cedar Grove hawk counts for the period 1952–1957 (Mueller and Berger, 1961: 175).

To reach the Cedar Grove Ornithological Station, according to Pettingill (1962: 45), ". . . drive eastward on U.S. Route 141 to the point near Cedar Grove where it separates from State Route 42. Continue east on Route 141, for about a half mile, until Route 141 turns north. Keep going east on a road, now gravel, which leads toward Lake Michigan. After a quarter mile, turn sharply north on the first road to the left and drive for about a half mile. This will cross Bahr Creek and take you into the sanctuary, where you can park the car. Nearby is a high bluff from which you can see the flights."

MILWAUKEE

According to Jung (1935) hawk migrations also are seen in Milwaukee along bluffs forming the waterfront. Many of these flights occur at considerable heights, but no additional information is available.

MINNESOTA

The general status of hawk migrations in Minnesota is documented by Hatch (1892) and by Roberts (1932). However, the best-known and most-detailed Minnesota hawk flights are those which occur at the western tip of Lake Superior at Duluth. The Broad-winged Hawk migrations which occur here in September are among the most impressive of any occuring on the North American continent.

DULUTH

In 1951, the Duluth Bird Club and the Minnesota Ornithologists' Union began a systematic study of autumn hawk migrations which occur along the western tip of Lake Superior (Hofslund, 1954, 1962, 1966; Green, 1962). A series of 600- to 800-foot bluffs rising from the shore of Lake Superior provide lookouts for hawk watchers. The Skyline Boulevard Lookout above 47th Avenue East is the chief spot. Recently the Duluth Audubon Society purchased the best lookouts and established the Hawk Ridge Nature Reserve —an exciting new wildlife sanctuary.

Most large Broad-wing flights occur in mid-September on days with westerly winds (Hofslund, 1954, 1962, 1966). Hawks usually are observed directly east or northeast of the Skyline Boulevard Lookout when first detected, but they are moving west to southwest

Observers on the Skyline Boulevard lookout at Hawk Ridge Nature Reserve, Duluth, Minn. This is a major concentration point for hawks migrating through the Great Lakes region. Photo by Donald S. Heintzelman.

as they pass over the lookout. Birds drifting along the shoreline turn
west in the area of Minnesota Point (Hofslund, 1966: 83). Hawk
counts at Duluth are impressive (table 11) as reported by Hofslund
(1966, personal communication) and Sundquist (1973).

To reach Hawk Ridge Nature Reserve in Duluth, travel east on
London Road (Highway 61 traverses London Road) to 47th Ave-
nue East, turn up the hill and continue for one mile to a stop sign
at Glenwood Street. Turn left and go about 0.8 mile to Skyline
Parkway. This is a dirt road turning sharply back to your right.
Continue on Skyline Parkway for one mile until the Hawk Ridge
Nature Reserve sign is reached. Trails lead to the lookouts (Henry
B. Roberts, letter of 7 November 1972).

UPPER MISSISSIPPI RIVER

During two days in November of 1972, along the bluff ridges
beside the Mississippi River along Route 61 in Wabasha and Win-
ona Counties, Reese (1973) noted a marked Bald Eagle migration.
A count of 147 individuals was obtained, and more may have passed.
Additionally small numbers of Goshawks, Red-tailed Hawks, and
Golden Eagles were seen.

5

New England Hawk Migrations

Large numbers of hawks migrate across New England during autumn, but relatively few systematic studies of hawk flights have been made. Some fragmentary records are published in *Records of New England Birds*. A few other major papers have appeared. Nonetheless, current knowledge of hawk flights in New England is not equal to that for New Jersey, Pennsylvania, and Tennessee, for example. Recent hawk migration studies in southern New England are coordinated by Hopkins and Mersereau (1971, 1972).

MAINE

SWAN ISLAND, KENNEBEC RIVER

On 9 September 1957, a migration of about 76 hawks, mainly Broad-wings, was seen crossing the Kennebec River between Swan Island and Dresden Neck (Powell, 1957). The birds were flying about 400 feet above the water and slowly moving due west. It is uncertain if these observations represented an isolated hawk flight or if Swan Island is near a concentration point for migrating hawks.

MOUNT CADILLAC, MOUNT DESERT ISLAND

No systematic hawk counts have been made from the summit of Mount Cadillac in Acadia National Park. However, fragmentary data indicate that this spot may be a good hawk lookout. On 30 September 1952, Howard Drinkwater (personal communication) counted 26 Sharp-shinned Hawks, 3 Cooper's Hawks, 2 Red-tailed Hawks, 2 Rough-legged Hawks, 2 Golden Eagles, 13 Marsh Hawks, 6 Ospreys, 1 Peregrine Falcon, 2 Pigeon Hawks, and 13 Sparrow Hawks from the summit of Mount Cadillac. Observations were made between 0900 and 1600 on a clear day with strong northwest winds and no cloud cover.

Drinkwater noted that the hawks approached from the northeast, crossed Frenchman's Bay, and picked up a strong reverse air current boiling up through a narrow, deep V-shaped gorge extending from sea level to the summit of the mountain near where he was

Frenchman's Bay as seen from Mount Cadillac, Mount Desert Island, Me. Photo courtesy of Acadia National Park.

stationed. These data suggest that Mount Cadillac may be a good hawk lookout.

NEW HAMPSHIRE

Limited hawk watching has been done in New Hampshire on a systematic, long-range basis, and only the southwestern quarter of the state has received moderate attention (Anonymous, 1953). Within recent years more attention has been devoted to hawk migrations, however (Robert W. Smart, letter of 23 October 1972).

LITTLE ROUND TOP

According to Robert W. Smart (letter of 23 October 1972), Little Round Top (Inspiration Point) near Bristol is one of New Hampshire's best hawk lookouts. The elevation of the site is 1,010 feet. This spot has been used mainly for studying Broad-wing flights in September. The best winds seem to be light and variable with an

The lookout at Little Round Top near Bristol, N.H. This is one of the state's better hawk lookouts. Photo by Elizabeth W. Phinney.

easterly trend. A strong northwesterly wind does not produce a good Broad-wing flight, but this is usually true at most lookouts. Broad-winged Hawks generally dislike flying when extremely strong winds prevail (Heintzelman and Armentano, 1964: 8).

Robert Smart provides the following directions for reaching Little Round Top: "From the center of Bristol start south towards Franklin on rt. 3A. As you are leaving the village there is a firehouse in the middle of a fork. Take the right hand fork, up a hill and then right at the next fork. Ignore all roads which are a sharp turn to the right. Follow the road as far as possible. This will take you to the Slim Baker Conservation Area and Day Camp. Park by the main building and follow the trail or tote road to the top of the hill, about a ten minute walk. The observation area is on the Northeast corner of the hill and gives superb views from due west through the north and the east to nearly due south. Only the southwest is blocked by trees and the hill is, therefore, useless during the spring flight." On 15 September 1971, 3,808 Broad-wings were counted here.

MOUNT BELKNAP

This lookout, elevation 2,378 feet, is located near Gilford. During September of 1963, the site was used for five days according to data filed at Patuxent Wildlife Research Center. Two Goshawks, 5 Sharp-shinned Hawks, 12 Cooper's Hawks, 6 Red-tailed Hawks, 5 Red-shouldered Hawks, 382 Broad-winged Hawks, 13 Ospreys, 3 Sparrow Hawks, and 32 unidentified hawks were counted.

NEW IPSWICH

Of several locations used during the early 1950's in an effort to locate good New Hampshire hawk lookouts and migration concentration points, only the field studies of Cora Wellman at New Ipswich produced reasonably large numbers of hawks (Wellman, 1957, unpublished data; Hill, 1957). The lookout is atop a 1,500-foot hill located four miles northwest of New Ipswich. Good views are present in all directions. Hawks migrate past the lookout in a southwestward direction after approaching from the north over Temple Mountain (Cora Wellman, letter of 29 September 1952 to

Chandler S. Robbins). Table 12 summarizes the New Ipswich hawk counts from data filed at Patuxent Wildlife Research Center.

UNCANOONUC MOUNTAIN

This site has been used for a few years (Robert W. Smart, letter of 23 October 1972). On 17 September 1970, over 2,500 Broad-winged Hawks were counted. The best flights develop on north or northwest winds. The birds generally approach from the northeast and continue toward the southwest. Trees restrict views at the site's parking lot, but a fire tower can sometimes be used to obtain a better vantage point. The elevation of this spot is 1,324 feet. Robert Smart provides directions: "From Main Street in Goffstown take Mountain Road at Esso Station. Stay on Mountain Road—at two intersections keep to the left around two sharp curves. Turn left again on Mountain Summit Road (the sign is usually up there but has a habit of disappearing). Stay on Mountain Summit Road all the way to the top—parking area near the tower."

VERMONT

Little detailed information is available regarding Vermont hawk migrations. However, filed at Patuxent Wildlife Research Center are hawk counts made by Thomas Foster on Hogback Mountain (elevation 2,300 feet) near Marlboro. On 29 September 1958, 4 Sharp-shinned Hawks, 1 Cooper's Hawk, 1 Red-tailed Hawk, 83 Broad-winged Hawks, 2 Ospreys, and 3 Sparrow Hawks were counted. On 13 September 1959, 1 Sharp-shinned Hawk, 1 Cooper's Hawk, 1 Red-tailed Hawk, and 72 Broad-winged Hawks were counted.

RHODE ISLAND

Nothing detailed has been written about hawk migrations in Rhode Island. Clement (1958: 118) states that John J. Lynch once observed a flock of 300 Broad-winged Hawks passing over Newport, and that the birds avoid water and move inland up the estuaries such as the Sakonnet River. Hanley (1968) makes a few other general statements without systematic supporting data.

MASSACHUSETTS

Although Massachusetts has many active bird watchers, only limited systematic hawk watching has been done in this state. Forbush (1927), for example, provides mostly general statements regarding Massachusetts hawk migrations. Bagg (1947, 1949, 1950) published some preliminary, but more detailed, information on hawk flights in the Connecticut Valley with particular emphasis on hawk watching at Mount Tom. In addition, some fragmentary records also are published in *Records of New England Birds* and in *American Birds,* but detailed yearly counts are unavailable.

MARTHA'S VINEYARD

According to Gus Ben David II (letter of 12 October 1973) of the Massachusetts Audubon Society, modest numbers of Peregrine Falcons regularly migrate over Martha's Vineyard during autumn. The bulk of the flight occurs between 30 September and 12 October. Most of the falcons arrive over the southeastern corner of Chappaquiddick Island, continue westward following the south shore of the Vineyard, then depart in a generally westward direction from the vicinity of Squibnocket Point and Zacks Cliffs. A few birds also are observed over Felix Neck from time to time during autumn. The geographic origin of these Peregrines is unknown, but Gus Ben David II suggests that they represent the tundra subspecies, *Falco peregrinus tundrius.*

MOUNT TOM

Goat Peak, in Mount Tom State Reservation, is the best-known hawk lookout in New England. Hawk counts were begun here soon after Hawk Mountain, Pennsylvania, was discovered (Hagar, 1937a, 1937b) and continue to the present (Bagg, 1970: 6–8). Hawk flights at Mount Tom begin during late August and continue through mid-November. This is similar to the migration season in Pennsylvania. The representation of species also is similar. The data in table 13 are filed at Patuxent Wildlife Research Center. This information is incomplete, but illustrates the importance of this site.

Hagar (1937a) states that Goat Peak provides a spectacular

The view upridge from Goat Peak, Mount Tom State Reservation, Mass. This is one of New England's most famous lookouts. Photo by Donald S. Heintzelman.

view of the Massachusetts part of the Connecticut Valley, Mount Greylock, the southern Green Mountains, Mount Monadnock, Wachusett, and Springfield, Hartford, and Meriden. On clear days the waters of Long Island Sound appear to the south in the distance. Nearer, the Connecticut River loops from side to side, finally passing through the gap between Mount Holyoke and Mount Nonotuck. Mount Sugarloaf and Mount Toby bound the northern edge of the Valley, and the famous Oxbow of the Connecticut River is in the same direction. Free Orchard Notch is to the west. The main part of the Tom Range appears in the southwest at an angle of 45 to 50 degrees. To the south, southeast, east, and northwest of Goat Peak the land falls away gently to the Connecticut. Most hawks appear along these wooded slopes.

CONNECTICUT

Although sizeable hawk migrations cross the state of Connecticut, Sage, Bishop, and Bliss (1913) present no detailed information.

The most detailed reports of Connecticut hawk migrations are those of Merriam (1877) and Trowbridge (1895). In addition, a moderate amount of unpublished data is filed at Patuxent Wildlife Research Center. Recently new interest developed in studying hawk migrations in Connecticut (Hopkins and Mersereau, 1971, 1972).

BALD PEAK

Donald Hopkins (personal communication) reports that this spot is one of Connecticut's better hawk lookouts. It is located in the northwest corner of the state about four miles from Salisbury. The Peak is in Riga Forest at an elevation of 2,010 feet. On 19 September 1971, observers counted 2,049 Broad-winged Hawks from this spot.

NEW FAIRFIELD

This site is north of Danbury at an elevation of 800 feet. Frances J. Gillotti and other observers made hawk counts here during the 1950's and 1960's, mainly in the Broad-wing season. The data in table 14 are filed at Patuxent Wildlife Research Center.

NEW HAVEN

Hawk flights passing along the shoreline of Long Island Sound in the vicinity of New Haven have been known for nearly a century. On 25 September 1875, Merriam (1877) observed a flock of 26 Red-tailed Hawks soaring overhead near New Haven. On 18 September 1886, Trowbridge (1895) began a systematic study of hawk migrations in the vicinity of New Haven. These studies continued for ten years and demonstrated that substantial hawk flights occur along the Sound, particularly on days when northerly winds prevail.

TALCOTT MOUNTAIN

This site is in Avon Township west of Hartford. During mid-September of 1953, the Rt. Rev. Robert M. Hatch spent a few hours on four days counting hawks from Talcott Mountain. His data, filed at Patuxent Wildlife Research Center, included 1 Red-tailed Hawk, 3 Red-shouldered Hawks, 79 Broad-winged Hawks, and 1 Sparrow Hawk.

Rev. Hatch (letter of 22 September 1953 to Chandler S. Robbins) stated: "I have been observing on Talcott Mountain for three years. Last spring I got several good counts there—nothing at all spectacular, but consistent counts. Ospreys were very noticeable during the spring flight, whereas I did not see a single Osprey there this September. It may be that this place is a much better spring flyway than an autumn one, although I have not been observing there long enough to prove this point. My records do show that the results of the fall of 1952 were not as consistent as those of last spring."

WESTPORT

During the 1950's and early 1960's hawk counts were made by Mortimer F. Brown from Overlook Road in Westport. This is located close to Long Island Sound. Apparently the hawks seen by Brown are birds using the same flyway discovered nearly a century earlier at New Haven by Trowbridge (1895). Table 15 is compiled from data filed at Patuxent Wildlife Research Center.

6

Middle Atlantic
Hawk Migrations

Within that area extending from Lake Erie to the Atlantic Ocean, and from southern New England to the southern states, are ridges and valleys, wind and water gaps, and highlands and coastal plains. Vast numbers of hawks migrate across this area during autumn, and hawk watchers enthusiastically gather on some of North America's best lookouts to observe these migrations.

NEW YORK

New York has thousands of active bird watchers, a long history of ornithological exploration, and several centers of ornithological study, but limited systematic field study has been devoted to the state's autumn hawk flights (cf. Bull, 1964, 1970; Eaton, 1910, 1914; Elliott, 1960; Eynon, 1941; Giraud, 1844; Pettingill and Hoyt, 1963; Ward, 1963).

DUTCHESS HILL

This site is located on Rose View Farm along Dutchess Hill Road near Hyde Park, New York. In September of 1972, moderately good hawk flights were seen here according to James W. Key (personal communication).

FIRE ISLAND

This long, narrow barrier island is located off the south shore of Long Island in Suffolk County. It is separated from the Oak Beach section of Jones Beach by the Fire Island inlet. Darrow (1963) apparently is the only observer who has made prolonged hawk counts here. The bulk of his work was done in the vicinity of Democrat Point at the western extremity of the island. This spot is 37.7 miles east-northeast of Sandy Hook, New Jersey.

After twelve years of observation, Darrow (1963: 7–8) demonstrated that Peregrine Falcons are the only raptors which consistently use Fire Island as a flight line. From Democrat Point the bearings on the lines of flight used by hawks and other birds ranged between 225 and 240 degrees with an average of 232 degrees. Peregrine Falcon flight lines never varied more than 232±4 degrees.

Large numbers of falcons are observed at Democrat Point when northwest winds occur. Weather conditions apparently are not as important to good hawk migrations along Fire Island as they are to inland ridge flights. Northwest winds are desirable, however. On 1 October 1960, for example, 546 Sparrow Hawks were observed at the inlet between 0900 and 1515, and a check later in the day showed that falcons were still migrating through the area, suggesting that as many as 700 Sparrow Hawks might have passed Fire Island that day.

Some accipiters, Marsh Hawks, and Ospreys also are seen, but few buteos appear at Democrat Point. Soaring hawks generally circle back toward the mainland, where more favorable air currents occur.

FISHERS ISLAND

Fishers Island is located at the eastern entrance to Long Island Sound. It serves as a connecting link between Rhode Island and Long Island, New York. Unlike the barrier islands along coastal New

York and New Jersey, Fishers Island was farmed until 1923 and is
a terminal morain with rock-covered hills (H. Lee Ferguson, Jr.,
letter of 4 May 1961 to Bertram G. Murray, Jr.).

Discovery of substantial autumn hawk migrations crossing the
island from east to west occurred in 1916, and gunners began intensive hawk shooting activities at that time (Ferguson and Ferguson,
1922; H. Lee Ferguson, Jr., letter of 11 August 1961 to Bertram G.
Murray, Jr.). The shooting continued until about 1924 or 1925,
when the eastern end of the island was developed. However, the
Fergusons continued their hawk shooting until the beginning of the
Second World War (H. Lee Ferguson, Jr., letter of 11 August 1961
to Bertram G. Murray, Jr.). During the early years large flights of
Sharp-shinned Hawks contained 1,000 or more birds (H. Lee Ferguson, Jr., letter of 11 August 1961 to Bertram G. Murray, Jr.), but by
1963 there was a drastic reduction in the numbers of Sharp-shins
seen (H. Lee Ferguson, Jr., letter of 22 May 1963 to Chandler S.
Robbins). Fishers Island is an autumn migration route. Few hawks
appear during spring (Ferguson and Ferguson, 1922: 496).

HOOK MOUNTAIN

Hook Mountain, elevation 750 feet, is located a few miles above
Nyack, Rockland County. It is a relatively short ridge running east
to west beside the Hudson River. The lookout, within sight of the
Tappan Zee Bridge three miles to the south, is a cleared ridge crest
about 75 feet long and 40 feet wide. It is reached by walking along
a trail marked with blue blazes from Route 9W to the flat mountain
summit. Views to the south, southeast, and west are unobstructed,
but some trees partly restrict visibility toward the east and north.

Autumn hawk counts were begun in 1966 by Mrs. Edward O.
Mills, who provided data for the period 1966 through 1970 (table
16). The 1971 data are from Mills and Mills (1971) and Thomas
(1971b). The Fyke Nature Association coordinated the Hook
Mountain hawk counts in 1971. Spring counts also are made here
(Thomas, 1971a). Hook Mountain data for 1972 are published by
Thomas (1973).

Observers on the lookout at Hook Mountain, N.Y. Photo by Donald S. Heint-zelman.

JONES BEACH

This Long Island site is a falcon flyway (Ward, 1958, 1960a, 1960b). The comparatively flat beach areas offer excellent views in all directions, but the best hawk-viewing area is around the amphitheater between the fishing station and Zachs Bay. Large numbers of small birds also congregate there during migration, and hawks apparently are attracted to these birds. Sparrow Hawks usually are the most abundant raptors. Buteos are rare. Sharp-shinned and Cooper's Hawks also appear in good numbers at times, and some winter along the strip.

Ward (1960a) reports on one excellent hawk flight: "On October 10, 1959, along with a good number of Sparrow, Pigeon, and Marsh Hawks, several Osprey, and two accipiters, it was my pleasure to see no less than 18 Peregrine Falcons—all this in one day between 7:30 a.m. and 1:30 p.m. Had I remained longer the total undoubtedly would have been much greater. Between October 2 and October 12 I observed the amazing total of 33 Peregrine Falcons at Jones Beach! This number represents only weekend birding I might add. I wonder what total a full week of observation would have produced."

On 13 October 1962, Ward (1963: 23) observed another excellent falcon flight. Between 0830 and 1500, he estimated that more than 300 Sparrow Hawks, 110 Pigeon Hawks, and 8 Peregrine Falcons, along with various other species, were seen. Sparrow Hawks are usually the most abundant autumn raptor migrants seen at this spot, however (Ward, 1960a).

Weather conditions favorable for Hawk Mountain and Cape May Point also are good for Jones Beach hawk flights, although falcons appear on any wind. Nevertheless, Ward (1960a) points out that northwest winds at 5 to 15 miles per hour prevailed on the day when he observed 18 Peregrines, whereas easterly winds occurred the following day and only 5 Peregrines were counted. One bird flew directly into the 20 mile-per-hour winds out over the ocean. Table 17 summarizes the results of hawk watching in 1958 at Jones Beach (Ward, 1960b).

MOUNT PETER

Hawk counting at Mount Peter began in 1958 and continues to the present (Baily, 1967, 1969; Rogers, 1971; Stiles Thomas, unpublished data). Baily (1967) described the site in detail: "Mt. Peter lookout is located on Bellvale Mountain . . . just off Route 17A north of Greenwood Lake, New York. The parking lot is on the north side of the road, on the western edge of the crest. From here the little wooden 'Summer House' on the crest itself is clearly visible. This is crowned by a wind vane, and during the hawk season we leave a thermometer and information about the watch (particularly a daily hawk list to date) in this shelter.

"Bellvale Mountain is a continuous razorback ridge running northeast to southwest, rising abruptly from the valleys on either side. There are actually four lookouts on this ridge. Mt. Peter itself has been described above. At Falcon Ridge, ½ mile up the ridge

Observers on the lookout at Mount Peter, N.Y. Photo by Donald S. Heintzelman.

where a gas pipeline cut crosses the thickly wooded ridge, the hawks are usually very low and closer than at Mt. Peter, but because of trees one can see only overhead, east and west, not up or down the ridge. A tower at this point would be ideal. Osprey Rocks, across 17A about ¼ mile down the ridge, is a row of rock outcroppings with a better view east than Mt. Peter but a very poor view west. It is practical only when the hawks are ridge flying on the east side or on a Broad-wing day. Osprey Rocks and Falcon Ridge are just off the Appalachian Trail. The fourth lookout, on the east side several miles farther up the Appalachian Trail near the origin of the ridge, is practical only on a Broad-wing day because of the limited view."

Hawk watchers drive to the top of the mountain, park near a restaurant, and walk about one hundred yards to the lookout. The Fyke Nature Association of Ramsey, New Jersey, coordinated the hawk counts at Mount Peter from 1958 through 1970. In 1971, and 1972 the Highlands Audubon Society of New Jersey assumed responsibility for conducting the counts (table 18).

THE TRAPPS

This spot, located at Mohonk Lake near New Paltz, is used as a hawk lookout for two days each autumn (Virginia Smiley, letter of 1 November 1970). Activities began in 1954 and continue to the present. Modest numbers of hawks have been reported in past years. The site is on the northern edge of the Shawangunk Mountains which, in New Jersey and Pennsylvania, are known as the Kittatinny Ridge.

PENNSYLVANIA

Pennsylvania is one of the most important states in the East for studying autumn hawk migrations. Among the major hawk counts being carried out in this state are those at Hawk Mountain Sanctuary, Bake Oven Knob, and Waggoner's Gap. There are also a number of other excellent lookouts in Pennsylvania which currently are inactive or are used irregularly.

BAKE OVEN KNOB

Bake Oven Knob is a State Game Land owned by the Pennsylvania Game Commission. It is located on the crest of the Kittatinny Ridge about twenty miles north of Allentown and sixteen miles northeast of Hawk Mountain Sanctuary. The Knob offers ideal hawk watching opportunities (Heintzelman, 1963a: 154–58): "At an elevation of 1600 feet, it occupies a position on the ridge at which the mountain makes a gentle semi-U-shaped curve. Hence the observer encounters a view over the whole mountain toward the east: a view quite similar to that seen from the Hawk Mountain Lookout. Detection of approaching hawks is therefore possible at great distances."

There are two lookouts at Bake Oven Knob. Wind direction determines which is used. When northerly and westerly winds prevail, the North Lookout (formerly called "The Point") is used. Since these are the predominating winds of autumn, most of each season's hawk count is made from this spot (Heintzelman and Armentano, 1964: 3). However, when easterly and southerly winds

The view northeast from the South Lookout at Bake Oven Knob, Pa. This lookout is used primarily when easterly and southerly winds prevail. Large numbers of hawks formerly were shot from this spot. Photo by Donald S. Heintzelman.

occur we use the South Lookout, which has an equally commanding view over the crest of the Kittatinny Ridge toward the northeast.

The history of Bake Oven Knob is similar to Hawk Mountain. Hawk shooters discovered the spot early in this century, but shooting apparently was not as serious as at Hawk Mountain until the latter was established as a wildlife sanctuary. Then shooting at the Knob increased greatly as gunners moved from Hawk Mountain to newly constructed stands at Bake Oven Knob.

Richard H. Pough was one of the first conservationists to visit the Knob. On 10 November 1935, he assisted Maurice Broun by timing an adult Bald Eagle which apparently passed Hawk Mountain twenty minutes later (Broun, 1939: 432). On 24 October 1936, Benjamin C. Hiatt (letter of 25 November 1968) began a series of visits to Bake Oven Knob.

"My wife and I, together with Joe and Lou Cadbury, would spend the morning at Hawk Mountain then, during the noontime

The North Lookout at Bake Oven Knob, Pa. This spot is used on days with prevailing northerly and westerly winds. Photo by Donald S. Heintzelman.

lull, move on to Bake Oven Knob. Since our visits were usually on Sunday, there was no active shooting. However, we were able to find many dead birds in the woods and heaped by the roadside. These included Red-tails (three with feet cut off), Broad-wings, Marsh Hawks, Ospreys, many Sharp-shins, a few Cooper's plus numbers of Blue Jays, Crows, and Flickers.

"We rarely met other birders . . . although we stayed at the South Lookout most of the time, or explored the woods below for specimens to skin and stuff.

"After an absence of several years I again visited Bake Oven Knob, beginning in 1951. This time I made trips alone or with my son, and on Saturdays, at which time the shooting stands were active. There was one on the north slope just opposite the old beacon tower, several spots on the south slope near the lookout and one large, well hidden one on the southwest edge of the rock pile.

"Gunners would use live pigeons tethered on poles to lure the hawks in close. Not only were the Accipiters attracted by these decoys, but other hawks as well. Very seldom did the protected birds escape gunfire, although there was ample time to make identification. The warden, Bill Moyer, was observed on several occasions . . . escorting gunners off the Knob after he had caught them with protected birds.

"I spent all of my time at the South Lookout or at the shooting stand further west along the ridge. Here the rocks had been heaped up to make a breastwork and comfortable shelter, large enough for four or five. The gunners didn't mind my presence and in fact would tell me where to look for birds they had shot. One injured Sharpie which I found was far from dead. She gave me quite a battle and was subdued only after I tossed my jacket over her head and wrapped her in it."

In 1936, Pennsylvania protected a few species of hawks, but law enforcement was difficult. The Game Commission made some efforts to protect hawks but was unable to cope with gunners participating in the activity. The shooting of protected birds continued.

On 18 September 1949, Clayton M. Hoff (personal communication) visited Bake Oven Knob and found about 20 gunners shooting hawks. Others were engaged in similar activities along the road and at locations west of the road. During a three-hour period prior to noon 380 shots were fired at passing hawks; 361 were illegally fired shots. The total of hawks shot at numbered 162 of which 154 were illegal targets (protected birds). Of the 38 hawks actually shot, 35 were protected. These included 30 Broad-winged Hawks, 3 Ospreys, 3 Sharp-shinned Hawks, and 2 unidentified hawks. Two passing Bald Eagles were not used as targets.

Hoff stated that "There was some display of sportsmanship on the part of some of these hunters in that they criticized some of their confederates for shooting at Ospreys but, at the same time, although they and I identified the approaching birds as Broad-wings, had no reluctance in using them as targets." One hunter commented of a fellow hunter: "He would shoot anything that flies, including his grandmother if she had wings." The shooting was done from improvised blinds using decoys (a live pigeon and a dead Broad-wing) in two instances. Passing hawks paid no attention to the decoys.

"Watching the casualties as they fell or glided," wrote Hoff, "it is my conclusion that not over one-third of the birds shot were killed, and that two-thirds were disabled, gliding into the woods where they probably died of starvation, becoming the prey of predatory animals, or dying from poisoning. Undoubtedly many of the hawks continue in flight with fatal injuries that might result in poisoning."

One final example illustrates the seriousness of the hawk shooting. On 22 September 1954, Nelson Hoy estimated that three or four times the 1,224 low-flying hawks which passed Hawk Mountain that day passed Bake Oven Knob. Several hunters shot many Broad-winged Hawks and some Ospreys. Most of the birds passed far out over the south valley and were missed at Hawk Mountain. The next day Lon Ellis and three other men visited Little Gap, another hawk-shooting spot a few miles up the ridge. They found about 75 freshly killed Broad-wings along with other hawks and a flicker (Broun, 1955). On his visits to Bake Oven Knob, Hoy salvaged some dead hawks and prepared them as study skins. Two are in the William Penn Memorial Museum collection, and many others remain in Hoy's private collection.

My first visit to Bake Oven Knob was in mid-October of 1956. In 1957, Hawk Mountain Sanctuary asked me to check shooting stands along the ridge for possible violations of a new law protect-

In Pennsylvania, prior to 1957, large numbers of migrating hawks were shot. These birds, all protected species, were shot illegally in 1956 near Bake Oven Knob, Pa. Included are eight Broad-winged Hawks, one Marsh Hawk (*lower left*), one Osprey (*center*), and one immature Peregrine Falcon (*lower right*). Photo by Donald S. Heintzelman.

ing hawks. I spent several hours counting hawks at the Knob on each of fourteen days. These are the first systematic hawk counts from the Knob. In 1961, yearly counts were begun (Heintzelman, 1963a, 1963b, 1966, 1968, 1969b, 1970d; Heintzelman and Armentano, 1964; Heintzelman and MacClay, 1972, 1973). Table 19 summarizes these counts.

To visit Bake Oven Knob, drive north on Route 309 for two miles from the junction of Routes 309 and 143 near New Tripoli, then turn right (east) onto a paved road. Continue for two miles, then turn left onto another narrow paved road running between a white house and several other buildings. Continue for about a quarter mile. At the point where the paved road turns sharply right, continue straight ahead onto a gravel road and follow this for about a mile to the top of the mountain. Park in one of two parking lots, and walk east on the Appalachian Trail for about one-third mile to the summit of the Knob. After crossing a boulder field and climbing a steep incline, look for an old cement foundation beside the trail. The South Lookout is located about 150 feet to the right of this spot. The North Lookout can be reached by walking along the Appalachian Trail for about a quarter mile more. After passing a small campsite, and walking along the north side of a large boulder pile,

climb the northeastern end of the boulder pile to the small, exposed
lookout.

BALD EAGLE FIRE TOWER

This tower is on Bald Eagle Mountain, elevation 1,800 feet,
about two miles northwest of Bellefonte, Centre County. During
September of 1958, according to data filed at Patuxent Wildlife
Research Center, Charles E. Trost spent four days (7, 8, 10 and 14
September) counting hawks migrating past the tower. About two
or three hours of observation were made each day. The following
birds were seen: 2 Sharp-shinned Hawks, 1 Cooper's Hawk, 10 Red-
tailed Hawks, 35 Broad-winged Hawks, 1 Bald Eagle, 1 Marsh
Hawk, 1 Osprey, 2 Sparrow Hawks, and 4 unidentified hawks.

BEAR ROCKS

This outcropping of boulders on the crest of the Kittatinny
Ridge is located about one and one-half miles southwest of Bake

The lookout at Bear Rocks, Pa. The summit of Bake Oven Knob appears in the
background. Photo by Donald S. Heintzelman.

Oven Knob. It provides a spectacular 300-degree view with the summit of Bake Oven Knob visible toward the northeast and excellent to very good views down both slopes of the ridge. Hawks, such as Sharp-shins, flying very low along the south slope might be missed by observers on Bear Rocks, however. Although little has been published about this spot, it is nevertheless an outstanding lookout. Alan and Paul Grout (personal communication) provided the unpublished counts in table 20.

A variety of spellings are used for Bear Rocks. Miller (1941) states that Bear Rock also is used. Poole (1932) indicates that Black Bears and Bobcats formerly may have lived among the huge boulders.

DELAWARE WATER GAP

On the west (Pennsylvania) side of Delaware Water Gap are located a series of rocky ledges along the Appalachian Trail. Just before the trail descends into the Gap, one enjoys an excellent view

The view northeast looking across Delaware Water Gap from the Pennsylvania side. Photo by Donald S. Heintzelman.

toward the northeast across the Gap to the New Jersey side. In September of 1972, I spent four and one-half days here making a preliminary hawk count. I recorded 1,235 hawks of eight species, including 1,177 Broad-wings (Heintzelman, 1973). Although this site probably is not a major lookout, it nonetheless has considerable value to hawk watchers, especially when easterly and southerly winds occur, and it is easily reached from the Mount Minsi Fire Tower overlooking Delaware Water Gap.

To reach the Delaware Water Gap lookout, drive to Tott's Gap, approaching from the north side (directions given later in the Tott's Gap section) and continue northeastward along the ridge crest road to the Mount Minsi Fire Tower. Then walk northeastward along the Appalachian Trail to the rocky ledges overlooking the Water Gap. Any convenient spot can be used as a lookout. The walk from the fire tower to the ledges requires about five minutes.

HAWK MOUNTAIN SANCTUARY

This is the best-known location in North America for observing autumn hawk migrations. The Hawk Mountain site was established in 1934 and now covers about 2,000 acres of privately owned forest bordering Berks and Schuylkill counties. It was the first wildlife sanctuary in the world specifically established to protect birds of prey (Broun, 1949a).

The reason for the initial establishment of the Sanctuary was to stop the shooting of migrating hawks which occurred there. This shooting, which reached considerable proportions by the late 1920's, probably began about 1915 although accurate information is vague. By 1925 it had become an extremely popular "sport" among local hunters (Poole, 1934: 18). A few naturalists, notably Henry W. Shoemaker, were aware of the shooting early in the 1920's, but George M. Sutton was the first ornithologist to visit the site in October of 1927 (Sutton, 1928; Poole, 1934; Broun, 1935). Sutton picked up many dead hawks and wrote a paper dealing with plumage differences, weights, and stomach contents of 158 birds of 4 species. On 27 October 1929, Earl L. Poole made his first visit to Hawk Mountain. For several years thereafter he took several boys with him to gather freshly killed hawks for preparation as study skins

Dead hawks slaughtered at Hawk Mountain, Pa., in October of 1932. The outrage of wildlife conservationists directed at this destruction of birds of prey led to the establishment of Hawk Mountain Sanctuary in 1934. Photo by Richard H. Pough.

for the Reading (Pennsylvania) Public Museum and Art Gallery (E. L. Poole, letter of 2 August 1969). During the autumn of 1931, and continuing through 1945, Dr. Poole spent many weekends on the lookout at Hawk Mountain.

The first effort to stop the hawk slaughter began in 1932. Richard H. Pough visited the mountain and described the shocking activities there (R. H. Pough, letter of 16 September 1969; Pough, 1932: 429–430; Broun, 1949a: 4–5). Pough's article also contained a letter sent to the Executive Secretary of the Pennsylvania Board of Game Commissioners requesting an investigation of "conditions on Blue Mountain [Hawk Mountain] with a view to insuring better enforcement of existing laws and, also, in the future to detail wardens to restrict hunters from the shooting of protected birds during this autumnal Hawk-flight."

In March 1933, Henry H. Collins, Jr., provided detailed de-

Observers on the North Lookout at Hawk Mountain, Pa. Prior to 1934, thousands of migrating hawks were shot from this spectacular vantage point on the crest of the Kittatinny Ridge. Elsewhere in Pennsylvania the slaughter of migrating hawks continued until 1957. Photo by Donald S. Heintzelman.

scriptions of the hawk slaughter at Hawk Mountain. Photographs taken by Richard Pough documented the article's information (Collins, 1933: 10–18). In October 1933, Richard H. Pough reported to a joint meeting of the National Association of Audubon Societies, the Hawk and Owl Society, and the Linnaean Society of New York that contact was made with real estate agents representing the owners of the mountain. In August 1934, Mrs. Charles Noel Edge secured a one-year lease on the mountain with an option to buy. Arrangements were made to have Maurice Broun serve as warden. He and his wife Irma arrived at Hawk Mountain early in September 1934 to begin their new and difficult duties. Broun (1949a) presented his experiences at Hawk Mountain during the early formative years of the development of the Sanctuary in *Hawks Aloft: The Story of Hawk Mountain,* to which readers are referred.

Hawk enthusiasts from around the world now come to Hawk

Mountain to enjoy watching the hawk migrations. Hawks may be observed from two main lookouts established for visitor use. The North Lookout, from which gunners shot hawks in pre-Sanctuary days, is on the crest of the Kittatinny Ridge. It provides a sweeping panorama of the Pennsylvania landscape. Hawks are seen approaching from a great distance as they ride updrafts or thermals along the mountain.

The newer second lookout is located several hundred feet behind the main entrance to the Sanctuary (the entrance has been moved several times since 1934). This South Lookout provides an excellent view overlooking the Great Valley and a geological formation called the River of Rocks but does not permit observers to see hawks flying along the north side of the ridge. Hence the South Lookout is of particular importance mainly when easterly or southerly winds occur. Many hawks tend to drift southward away from the main ridge under those conditions. Often they are missed by observers on the North Lookout as they fly across The Kettle.

The South Lookout at Hawk Mountain, Pa. This spot is particularly valuable when easterly and southerly winds prevail. Photo by Donald S. Heintzelman.

Table 21 summarizes the yearly (autumn) hawk counts made at Hawk Mountain. Data are taken from *Hawks Aloft* (Broun, 1949a) and from more recent *News Letters to Members* issued yearly by the Hawk Mountain Sanctuary Association.

To visit Hawk Mountain, drive to either Kempton or to Drehersville in eastern Pennsylvania (both are located a few miles from Hamburg) and follow the signs pointing to the Sanctuary.

LARKSVILLE MOUNTAIN

This site, located in Luzerne County west of Wilkes-Barre at an elevation of 1,604 feet, had been a hawk-shooting spot for decades. In 1958, limited observations were made here by Harry Brown, Edwin Johnson, and William Reid, whose unpublished data are filed at Patuxent Wildlife Research Center. These data, and the hawk shooting which continued for many years at this spot, indicate that Larksville Mountain is an adequate autumn hawk lookout. Table 22 summarizes the 1958 hawk counts.

LEHIGH FURNACE GAP

This wind gap * on the crest of the Kittatinny Ridge is located about one mile northeast of Bake Oven Knob. With prevailing northwest winds, watchers observe hawks from a rocky outcropping along the north slope of the ridge approximately one mile northeast of the dirt road which crosses the mountain. This spot is reached by walking along the Appalachian Trail until the rocks are seen, then following an obscure trail through scrub oak to the north side of the ridge. On days with prevailing easterly or southerly winds, some hawks can be seen from the electrical power line on the crest of the ridge beside the road. This spot has limited value, however. Lehigh Furnace Gap currently is used mainly by hawk banders (see chapter 1).

LEHIGH GAP

Refer to chapter 1 for information regarding this spot as a hawk banding station.

---

* Gaps or notches in mountains cut by streams which later were diverted to other places. They are common geologic features in the Appalachians.

LITTLE GAP

Little Gap is a former hawk shooting stand on the crest of the Kittatinny Ridge just northwest of Danielsville, Northampton County. Several spots on both sides of the road crossing the mountain are suitable as hawk lookouts. They are reached by walking along the Appalachian Trail. One spot is reached by walking westward along the trail to a pipeline. Another is a rocky outcropping east of the road.

THE PINNACLE

This is the highest peak in Berks County (elevation 1,620 feet). It is located about five miles south of Hawk Mountain Sanctuary and is visible from the Sanctuary lookouts. Although difficult to reach, the Pinnacle's summit is a rewarding lookout, particularly when easterly and southerly winds occur. Hawks drifting south from the Kittatinny Ridge head across part of the Great Valley toward the Pinnacle, then follow its spur of the Kittatinny Ridge southwestward. Details regarding trails leading to the summit of the Pinnacle can be secured from Hawk Mountain Sanctuary.

The Pinnacle, Berks County, Pa. Although difficult to reach, it is a good hawk lookout on days with easterly and southerly winds. Photo by Donald S. Heintzelman.

PIPERSVILLE

Pipersville is a small town in central Bucks County a few miles
west of the Delaware River. Observations made by Ann Webster
demonstrate that a considerable number of hawks drift across this
area during autumn despite the site's position roughly halfway
between the Kittatinny Ridge and coastal New Jersey. Hawks
crossing Pipersville apparently are engaged in cross-country migrations, but the nearby Delaware River and several minor ridges may
have some influence upon Ospreys and other species. Broad-winged
Hawks observed at this spot almost certainly are reflecting this species' use of thermals and the tendency of these bubbles of warm
air to drift cross-country, however. Table 23 summarizes the Pipersville hawk counts based upon data provided by Ann Webster (personal communication).

THE PULPIT (TUSCARORA MOUNTAIN)

This site, elevation 2,175 feet, is located on the summit of Tuscarora Mountain. It is reached by driving west on Route 30 from
Chambersburg toward McConnellsburg. At the top of the mountain,
walk to the rear of the inn and follow the sign marked for the Pulpit.
It is a flat area of soil on top of large rocks. A U-shaped pile of rocks,
opening toward the south, is located on the flat area.

Views toward the north, west, and southwest are good, but
the view toward the east is limited because the ridge fails to drop
away as sharply as it does in the other directions. Hawk counts
were begun in 1967 under the direction of Carl L. Garner, who provided the data in table 24, and who continues to coordinate hawk
watching activities at this spot.

ROUTE 183

This spot is located on the crest of the Kittatinny Ridge at the
point where the road crosses the mountain north of Strausstown,
Berks County. Hawk watching is done on days with westerly or
northerly winds. Observations are made from the porch of a cabin
beside the Appalachian Trail.

Observers on the Pulpit, Tuscarora Mountain, Pa. Photo by Dorothy Beatty.

Table 25 is compiled from data provided by Earl L. Poole, who spent a few hours of observation at this site on various days. Counts were not made during 1967.

STERRETT'S GAP

During the late 1930's and early 1940's, Sterrett's Gap was used as a hawk lookout by Edward Snively Frey (1940: 247–250; 1943). Frey (1940: 248) described the site as "a shallow wind gap on the Kittatinny Ridge between Cumberland and Perry Counties twelve miles west-southwest of the Susquehanna River and 'down' the ridge as the birds fly, seventy miles from Hawk Mountain. This is a region of broken and converging ridges immediately to the north of the Kittatinny and of two companion ridges just to the north and paralleling the Kittatinny; and in close proximity to it here but east of the river they begin turning away from it obliquely in a northeastward direction, gradually at first and at last more sharply until at a point due north of Hawk Mountain they

are some miles distant. In all this region, the Kittatinny, as elsewhere along its course, is the only continuous ridge flanked on the
south by the Kittatinny Vale, a broad valley, and on the north, more
and more as it draws nearer to Sterrett's Gap, by numerous ridges
and tumbling hills which are the beginnings of the Appalachian
Plateau."

Frey's efforts covered only a few years, but his studies produced important information from an excellent site which gives us
valuable comparative insights into the autumn migration routes
used by hawks in southcentral Pennsylvania. Table 26 summarizes
Frey's (1943: 54) Sterrett's Gap hawk counts for the period 1938
through 1941.

STONY CREEK FIRE TOWER

This tower is on the crest of Stony Mountain a few miles north
of Harrisburg, Dauphin County. On 8 September 1966, I counted
5 Turkey Vultures, 1 Red-tailed Hawk, and 2 Broad-winged Hawks
during two hours of observation on the tower. Moderate northwest
winds prevailed, and visibility was good. The tower probably is not
an important hawk lookout.

TOTT'S GAP

This small wind gap is located on the Kittatinny Ridge on the
border between Monroe and Northampton counties. It is reached
from Delaware Water Gap, Pennsylvania, via the Cherry Valley
Road to a golf course, then west on the Poplar Valley Road to the
Tott's Gap Road. Follow the latter to the top of the mountain.
There turn left (northeast) onto the gravel road running along the
ridge crest to the Mount Minsi Fire Tower several miles distant.
Soon after turning onto the ridge crest road, note a large communications tower. About one-quarter mile beyond this, note a radio
tower and cleared right-of-way for a pipeline. This is the Tott's
Gap hawk lookout. Remain near the tower on days with northerly
or westerly winds, but walk to the south side of the ridge along
the pipeline when easterly or southerly winds occur.

The view looking north from Tott's Gap, Pa. Photo by Donald S. Heintzelman.

The hawk counts in table 27 are based upon unpublished data provided by Howard Drinkwater (personal communication) and upon more recent data gathered by Heintzelman (1973). Tott's Gap has limited value as a hawk lookout, although most birds seen here probably are entering Pennsylvania from nearby New Jersey.

TRI-COUNTY CORNERS

This is an important hawk trapping and banding station. See chapter 1.

TUSSEY FIRE TOWER

This tower, elevation 2,220 feet, is located on Tussey Mountain in Huntington County. Wood (1967: 24–25) recommends it as a lookout. The tower is reached by an unpaved road from State Route 26 in Pine Grove Mills Gap.

WAGGONER'S GAP

This well-known lookout (also spelled Wagner's Gap) is one of the more important hawk watching sites in central Pennsylvania west of the Susquehanna River (Carlson, 1966: 163–165; D. L. Knohr, letter of 27 August 1969). It is located in Cumberland County about twelve miles northwest of Carlisle at a point where Route 74 crosses the mountain into Perry County.

A radio relay tower is located on one side of the road at the

Observers on the lookout at Waggoner's Gap, Pa. The Kittatinny Ridge ends a few miles southwest of this spot. Photo by Donald S. Heintzelman.

gap, and a large parking lot is across the road. When southerly and easterly winds occur, hawk watchers remain near the tower and look northeastward along the south slope of the ridge and across the valley. When westerly and northerly winds prevail, observers follow a well-marked path around the parking lot and continue to the top of a large boulder pile on the ridge crest. This spot provides good views toward the west and northwest. Hawks appear overhead at low altitudes and are relatively easy to identify and count.

Systematic hawk counts were begun in the early 1950's by D. L. Knohr, who provided most of the data summarized in table 28. Limited additional data for the very early years also were filed at Patuxent Wildlife Research Center. Virtually none of these data previously have been published.

NEW JERSEY

A great deal of hawk watching has been done in New Jersey since the late 1930's, and the results indicate that it is a very important state in which to study hawk migrations. During recent years three major lookouts have been used systematically: the Montclair Hawk Lookout Sanctuary in northern New Jersey, Cape May Point at the southern tip of the state, and Raccoon Ridge on the crest of the Kittatinny Ridge near Delaware Water Gap. In addition to these three sites, numerous other minor lookouts also have been described by Edwards (1939), Lang (1943), and Heintzelman (1972e). The most complete yearly record of hawk counts in New Jersey is that published for the Montclair Hawk Lookout Sanctuary.

BEACH HAVEN

Edwards (1939) mentions this location as a coastal flyway used primarily by migrating falcons during both spring and autumn. The birds are most readily seen from the outer beaches where they fly low, at times barely clearing the dunes. No hawk counts are available from this site, however.

BEARFORT MOUNTAIN

This is not a major hawk lookout, but it nevertheless is of value as a secondary observation point. Moderate numbers of hawks pass the spot. Urner Ornithological Club members (Newark, New Jersey) visit Bearfort Mountain fairly regularly to watch migrating hawks. The lookout is an outcropping of rocks beneath the Bearfort Fire Tower in West Milford Township, Passaic County. Koebel (1970) describes the site: "Bearfort Tower is reached by taking Rte. 23 north to Union Valley Road, one mile past Newfoundland. Take a right on Union Valley Road and travel five miles north to Stephens Road. Go left on Stephens Road for 7/10 of a mile to a foot path on your left. 30 yards beyond the footpath, there is an area where cars may be parked under some Hemlock trees. Follow the foot path up the mountain for about one half mile. The lookout rocks are located off the right side of the trail just short of the tower.

"Once on the rocks, your short hike is rewarded with one of the most beautiful views in the state. Laid out below you is one of the few wilderness areas left in New Jersey: Cedar Pond Valley, with a mixed deciduous woods, some stands of evergreen, rock ledges, impenetrable bogs and a lake. The lookout (c. 1,300 ft.) affords a commanding view to the northeast, north, west and southwest, with part of Wawayanda Ridge enclosing the Valley on the far west side."

Bearfort Fire Tower was used as a lookout in the late 1930's, then was abandoned until 1967, when Harold Rae and Scott Moorhouse spent some time there late in the season. It was used again in October of 1968 by Moorhouse, Irving Black, and others. Koebel (1970: 10–11) states that hawks pass the Bearfort lookout at eyelevel or below, providing wonderful views of the plumage of the birds. Northwesterly winds create the best flights. Table 29 summarizes Koebel's (1970) data.

BOWLING GREEN FIRE TOWER

The Bowling Green Fire Tower is located in the highlands of New Jersey. Spring and autumn hawk flights occur here (Edwards,

The Bearfort Fire Tower, N.J. Photo courtesy of the Newark Museum.

1939), but the flights are scattered in comparison with those along the Kittatinny Ridge. This tower probably is a minor lookout. On 8 September 1969, Irving Black (personal communication) and Scott Moorhouse counted 1 Red-tailed Hawk, 10 Broad-winged Hawks, 1 Marsh Hawk, 4 Ospreys, 5 Sparrow Hawks, and 2 unidentified hawks from the tower.

BREAKNECK MOUNTAIN

This is another minor observation point in the highlands of northern New Jersey (Edwards, 1939). Years ago it was a hawk shooting stand, but it has not been used to any extent recently.

CAPE MAY POINT

Cape May Point is the most famous hawk lookout in New Jersey and one of the most famous in North America. Its importance as a major autumn concentration area for migrating birds is well documented by Witmer Stone (1937) in *Bird Studies at Old Cape May*.

The history of autumn hawk migrations at Cape May Point involves a tremendous amount of hawk shooting. Stone (1937: 265) stated that the shoots probably began prior to 1885, but the slaughter reached major proportions by 1920. During one week in September of that year at least 1,400 Sharp-shinned Hawks were killed (Stone, 1922: 567). In 1931, the National Association of Audubon Societies arranged to have a state game warden and a special warden, George B. Saunders, on duty at the Cape to prevent protected hawks from being shot. The effort was reasonably successful. Of an estimated 10,000 hawks passing the Cape, 925 were killed. In 1932, Robert Porter Allen was the special warden on duty. Of 5,765 hawks counted by Allen, 366 were shot. In 1935, another special warden, William J. Rusling, reported 1,080 hawks killed out of 13,452 observed (Stone, 1937: 266). The statistics in table 30, from Rusling's 1935 records (Stone, 1937: 266–267), illustrate the seriousness of the former hawk shoots at Cape May Point.

The first systematic hawk counting at Cape May Point was done in the 1930's by Allen and Peterson (1936), who worked out

The famous lighthouse at Cape May Point, N.J. On northwest winds large concentrations of hawks and songbirds occur here during autumn. Photo by Donald S. Heintzelman.

a timetable of hawk flights, listed the order of abundance of the species, and considered the paths followed by migrants in relation to wind velocity and direction. Information on food habits of migrating hawks, based upon stomach examination of dead birds, also was secured. Superficial hawk count comparisons also were made with other hawk lookouts.

Since 1935, many individuals have birded at Cape May Point, but few carefully planned hawk counts have been made. In 1965, Ernest A. Choate made a seasonal count, and this was repeated in 1970 (Choate and Tilly, 1973). As an adjunct to hawk trapping and banding at the Cape in recent years, hawk counts again are being made by the banders (Clark, 1968, 1969, 1970, 1971, 1972).

Because of the geography of Cape May Point, and the fact that flights are concentrated in different areas on some days and widespread on others, it is difficult to count hawks without having some duplication in the counts. Thus the Cape is not an ideal place to make quantitative hawk migration studies. Nonetheless, information from this spot provides insights regarding general migration patterns and related phenomena. Moreover, Cape May Point is a major concentration point for falcons during autumn. Table 31 summarizes the hawk counts from this spot based upon information reported by Allen and Peterson (1936), Choate (*in* Heintzelman, 1970e), Choate (1972), Choate and Tilly (1973), and Clark (1972, 1973).

CATFISH FIRE TOWER

This tower on a side spur of the Kittatinny Ridge is located about five miles northeast of Raccoon Ridge. It is reached by driving along the Blairstown-Millbrook road to the point where the Appalachian Trail crosses the highway about two miles southwest of Millbrook, then walking one and one-half miles southwest along the Appalachian Trail to the tower.

Hawk counts made here during 1970 and 1971 suggest that the site is most effective on southerly or easterly winds (Heintzelman, 1972a, 1972b). Many birds known to pass through the vicinity on days with northerly or westerly winds are missed by observers on the tower (Heintzelman, 1972a, 1972b; Tilly, 1972a, 1972b). Table 32 summarizes the 1970 and 1971 hawk counts from this spot.

CULVERS GAP

This site occasionally is used by hawk watchers. It is in Sussex County overlooking the northeastern edge of Culvers Lake. Hawks seen here are part of the flights which use the Kittatinny Ridge for varying distances. On 13, 14, and 20 September 1958, Howard Drinkwater (personal communication) counted the following hawks here: 11 Sharp-shinned Hawks, 1 Cooper's Hawk, 4 Red-tailed Hawks, 68 Broad-winged Hawks, 5 Marsh Hawks, 13 Ospreys, and 1 Sparrow Hawk. On 12 September 1959, 1 Sharp-shinned Hawk, 1 Red-tailed Hawk, 18 Broad-winged Hawks, 3 Marsh Hawks, and 2 Sparrow Hawks were counted.

HIGH MOUNTAIN

Located northwest of Paterson, Passaic County, High Mountain is part of the Watchung Mountains (Edwards, 1939). This is a fishhook-shaped, 35-mile-long series of ridges between Paterson and Somerville (Lang, 1943: 347, 351). The most recent hawk counts at this site were made by Irving H. Black (letter of 16 October 1969) during 1965, 1966, and 1967 (table 33). The data were collected during September except for 1 October 1967. High Mountain is useful mainly as a comparative lookout.

MONTCLAIR HAWK LOOKOUT SANCTUARY

This Sanctuary is one of three hawk lookouts in New Jersey used annually. Breck (1960b) described the early use of this site. From 1937 through 1942, members of the Urner Ornithological Club used this spot regularly during spring and autumn. Prior to that, before the town of Montclair prohibited the use of firearms, gunners shot hawks here. Since 1957, systematic hawk counts have been made every September with particular emphasis on Broadwing migrations. In 1959 the Montclair Bird Club purchased the site and gave it to the New Jersey Audubon Society.

Observers on the lookout at the Montclair Hawk Lookout Sanctuary, N.J.
Hawks were shot here earlier in this century. Photo by Donald S. Heintzelman.

Table 34 summarizes the Montclair hawk counts based upon
data published by Redmond and Breck (1961), Breck (1962),
Breck (1963), Breck and Breck (1964), Breck and Breck (1965),
Breck and Breck (1966), and Bihun (1967, 1968, 1969, 1970, 1972,
1973). Additional raw data are contained in mimeographed reports
issued yearly by the Montclair Bird Club, and in earlier mimeographed reports written by George Breck.

The Montclair Hawk Lookout Sanctuary is reached from Upper
Montclair by driving west on Bellevue Avenue to Upper Mountain
Avenue. Turn right and continue for one-quarter mile to Bradford
Avenue. Then turn left and go one-quarter mile to Edgecliff Road,
which is the second street on the right. Continue to the top of the
hill and one-quarter mile beyond an old rock quarry to the parking
space at the corner of Crestmont Road. Walk along a gentle slope
on the south side of the street to the top of the quarry, from which
hawk watching is done.

RACCOON RIDGE

This is New Jersey's most important ridge lookout. It is on the crest of the Kittatinny Ridge overlooking the upper end of Tocks Island and the Delaware Water Gap National Recreation Area. There are two lookouts available. The Upper Lookout formerly was the site of a fire tower. One can still see a few steel supports on the bare, windswept ridge crest. From here one has superb views over the Kittatinny Ridge toward the northeast, over the Delaware River flowing along the base of the north side of the mountain, and slightly restricted views over the south side of the mountain with the lower pond of the Yards Creek Pump Storage facility visible. The Lower Lookout is situated about two city blocks northeast of the Upper Lookout and is similar to the latter.

The importance of Raccoon Ridge as a hawk lookout was discovered in the late 1930's (Edwards, 1939). The name was invented to prevent gunners from discovering the site and using it as a shooting stand. Many hawk watchers have visited the spot over the years, but few detailed hawk counts have been published. Table 35 is a composite summary of information gleaned from the unpublished field notes of William Rusling for the years 1939 to 1941, from

The view north-northeast from Raccoon Ridge, N.J. Photo by Fred Tilly.

Howard Drinkwater's notes for 1951, 1952, and 1959, from the notes
of Drinkwater and Clarence D. Brown for 1954, from the notes of
Drinkwater and Fred Tilly for 1968, and from Fred Tilly's notes for
1969 and 1970. During the autumn of 1971 and 1972, Tilly (1972a,
1972b, 1973) made the first season-long hawk counts at Raccoon
Ridge.

This spot is difficult to reach. One approach is by driving to
the Yards Creek Pump Storage Station near Blairstown, Warren
County. After entering through the main gate, drive to a picnic site
and park. Then walk up a paved road to the point where a power
line runs down a small valley and up and over the top of the ridge.
Follow this very steep power-line cut to the top of the Kittatinny
Ridge, then walk northeastward along the Appalachian Trail to the
Upper or Lower Lookouts.

STAG LAKE

During the early 1900's Justus von Lengerke shot large numbers of hawks from a site known as Stag Lake in Sussex County (von
Lengerke, 1908; Burns, 1911: 234–235; Broun, 1949a: 6). Many of
these birds are preserved as study skins in the American Museum
of Natural History (Broun, 1949a: 6; Amadon, 1967: 54). Currently,
von Lengerke's shooting site is not used by hawk watchers.

SUNFISH POND

This glacial lake is located on the crest of the Kittatinny Ridge
about six miles northeast of Delaware Water Gap. It is reached via
a foot trail leading from the Yards Creek Pump Storage Station, or
via various entrance points onto the Appalachian Trail crossing
the Delaware Water Gap National Recreation Area. Information
for reaching Sunfish Pond is available at the information centers in
the National Recreation Area. Little hawk watching has been done
at Sunfish Pond.

SUNRISE MOUNTAIN

Sunrise Mountain, elevation 1,653 feet, is located in Stokes
State Forest, Sussex County. It is an important lookout used irregu-

larly by various hawk watchers. Birds passing this site form part of the general raptor movements using parts of the Kittatinny Ridge. Table 36 is compiled from data provided by Howard Drinkwater and Fred Tilly, and from data filed at Patuxent Wildlife Research Center.

DELAWARE

Mohr (1969: 5–6, 13) provides the best summary of the hawk watching done in Delaware. "Northern Delaware lies almost precisely between the two world-famous hawk lookouts [Hawk Mountain and Cape May Point] and a brisk morning between September 10 and October 5, can bring exciting flights—scores or hundreds of Broadwings, sometimes at tree-top level at Brandywine Creek State Park; other times appearing suddenly over downtown Wilmington, and circling over the office buildings for many minutes.

"At the Hawk Mountain lookout, and at nearby Bake Oven Knob, the hawks can be seen riding the updrafts that sweep the ridges of the Appalachians for a thousand miles and more. Along the New Jersey Coast the hawks follow the shoreline southward and converge over Cape May Point. Broadwings and other buteos circle upward, until they are out of sight. Small species, apparently fearful of the eleven-mile stretch of open water, will *backtrack* northward along Delaware Bay, to a narrower crossing up river. Efforts to pinpoint the crossing area have been unproductive. By either route, the hawks mysteriously disappear. They have consistently eluded watchers stationed on the Delaware side of the Bay.

"The birds seen in northern Delaware—over the city, over Fairfax, Hoopes Reservoir, and Brandywine Creek State Park—probably represent a more scattered population, only a small proportion of the hawks moving southward from more northern nesting territories. The Broadwings, which fly to South America for the winter, are a common nesting bird in northeastern forests, including Delaware.

"The valley of the Brandywine, especially where flanked by ridges, evidently attracts the migrants. Updrafts or thermals give the slower flyers a chance to catch up to the more skilled gliders. Appearing a few at a time, the hawks may spend five or ten minutes regrouping into a fairly tight flock as they ride a mile or more before peeling off into a narrow flight pattern and disappearing swiftly.

"Oddly enough, almost all of these hawks deviate from their south or southwestern flight line as they reach the sharp western bend in the Brandywine at Rockford Park, at the northern edge of Wilmington. Abruptly, they turn west—usually at great altitudes—possibly heading directly for the Blue Ridge, well to the west of York, Pennsylvania, or to Havre de Grace, Maryland, where big flights have been reported."

MARYLAND

A number of people have participated in autumn hawk counts at various Maryland locations since the late 1940's. The first organized hawk count occurred on 17 September 1949, when thirty-eight members of the Maryland Ornithological Society manned observation posts on the parallel ridges in Frederick, Washington, Allegany, and eastern Garrett counties. Table 37 summarizes the results of this effort (Robbins, 1950: 2–11). Of the twenty-one observation sites used, many were fire towers. Eight sites reported 200 or more hawks, and one (Wills Mountain) had over 1,000 hawks during one and one-half hours of observation. Only stations reporting 200 or more hawks are included in table 37. Consult Robbins (1950) for full details. The appearance of a Rough-legged Hawk in September is doubtful and later was rejected by Stewart and Robbins (1958).

ASSATEAGUE ISLAND

This is a narrow barrier island one mile wide and thirty-five miles long. The northern two-thirds are in Maryland and the southern one-third is in Virginia (Berry, 1971). The site has been known as a Peregrine Falcon flyway since 1938, when falconers began trapping these birds during autumn. Until recently little information was available on these flights, however. Nonetheless, unpublished data compiled by J. L. Ruos (personal communication) and data published by Berry (1971) and Ward and Berry (1972) provide an index to the island's Peregrine Falcon flights (table 38).

COVE POINT

This site is near the point where the Patuxent River flows into Chesapeake Bay. In 1949 large hawk flights were seen here. Robbins (1950) reported that "The highest one-day hawk count ever reported for Maryland was made on September 21 at Lore's Pond just north of Solomons in Calvert County by George Kelly. After receiving word of a large flight on the previous day, Mr. Kelly stood watch from 10 to 5 and counted 2,214 hawks. Eleven hundred of these were sighted at 10:25 as they passed high overhead in a single long flock. As early as 10 o'clock a flight of 96 was seen, and by 11:15 when the wind shifted from northwest to southwest, 2,187 had been counted." Kelly later observed hawks arriving on a broad front moving just south of west from Chesapeake Bay, from Cove Point to Drum Point. The birds were not concentrated at a single spot. The largest flights developed with light northwesterly or northeasterly winds. Table 39 summarizes the Cove Point hawk counts (Robbins, 1950).

MONUMENT KNOB

Beaton (1951: 166–168) and Carlson (1966: 165–166) consider Monument Knob an important hawk lookout. Its elevation is 1,500 feet, and it is located in Washington Monument State Park on South Mountain near Boonsboro (Heintzelman, 1972d). The monument is a stone tower from which hawk watchers have a view down the west slope of the ridge and over the valley. Accipitrine hawks and falcons generally fly just over the trees about halfway down the slope. Hawks approaching from the north follow a line slightly west of the crest of the ridge and can be spotted in the distance and followed as they fly overhead. The view to the north is partly blocked by trees. Table 40 is based upon data published by Beaton (1951) and unpublished data provided by Herbert E. Douglas (personal communication).

WHITE MARSH

Autumn hawk migrations have been studied at White Marsh for a number of years (Hackman, 1954; Hackman and Henny,

1971). This rural area is located about fourteen miles northeast of Baltimore. Its gently rolling land on the Piedmont Plateau reaches an elevation of 180 feet above sea level. Unlike hawk flights along the mountain ridges, White Marsh flights usually occur on south-west winds with Broad-wings forming the bulk of the September migrations (Hackman, 1954). Other species occur from mid-August through mid-November.

The hawk flights at White Marsh are modest in size, making this spot a relatively minor location. Hence the use of White Marsh data to evaluate raptor trends in the northeast (Hackman and Henny, 1971) is a questionable technique. Certainly there is no reason to believe that a minor site such as White Marsh is more typical of raptor population changes, in terms of its hawk counts, than are much more important lookouts such as Hawk Mountain.

VIRGINIA

Aside from Rusling's (1936) detailed investigation of Cape Charles hawk flights, only limited hawk watching has been done inland along Virginia's mountain ridges. Among the fire towers used as hawk lookouts on the Shenandoah Mountain are Reddish Knob, Meadow Knob, and High Knob (Carpenter, 1949, 1960). These sites are in a straight line about six or seven miles apart (Carpenter, 1960). Table 41 summarizes combined counts from these towers (Carpenter, 1960).

CAPE CHARLES

From 22 September through 11 November 1936, William J. Rusling made a detailed study of hawk migrations in the vicinity of Cape Charles. The investigation was made at the request of Richard H. Pough, then of the National Association of Audubon Societies. Rusling's (1936) observations were presented in a comprehensive report which contains the most detailed analysis yet made of Cape Charles hawk migrations. Through the cooperation of Frederick R. Scott of the Virginia Society of Ornithology, a copy of the report was made available to me. Table 42 is based upon this document.

Cape Charles lies about twelve miles north of the large peninsula called Delmarva, which is composed of Delaware and portions of Maryland and Virginia. The name Cape Charles also is applied to the end of the peninsula, about one mile south of the town of Kiptopeke. Most of Rusling's hawk counts were made from a low sandy point extending for a few hundred yards into the mouth of Chesapeake Bay. This site, which Rusling called "The Point," was located at the end of the peninsula below Kiptopeke.

Hawk flights at this spot were known to generations of people at the time Rusling worked in the area. No previous study had been made of the flights, however.

Rusling (1936: 11) noted that "A northeast wind blowing over the coastal area always brings large numbers of migrants to Cape Charles. Such a wind aids the birds as they fly over the coastal area, bringing them into the neck of the peninsula from where they fly south over the land and arrive south of the little town of Kiptopeke in great concentrations. Thus the northern part of the peninsula witnesses flights on a northwest wind while the southern end gets them on a northeast wind."

Frederick R. Scott (letter of 19 January 1970) suggests that changes have occurred in the species composition of the hawk flights in recent years. Bald Eagles almost never are seen, and Cooper's Hawks, Pigeon Hawks, and Peregrine Falcons are much less common now.

GREAT NORTH MOUNTAIN

This mountain rises south of Mayfield and runs southward along the Virginia-West Virginia border. DeGarmo (1953: 41) states that the fire tower about four miles from the northern end of the mountain in Virginia is the only spot for observing both slopes of the ridge. Limited use in 1952 suggested that Great North Mountain could be a favorable lookout.

HIGH KNOB

This Knob, on Shenandoah Mountain, lies in Rockingham County within sight of Meadow Knob. These stations are less than three miles apart (Carpenter, 1960). Yearly hawk counts are unavailable for this site, but some combined counts for the period 1949–1959 are published (Carpenter, 1960).

HIGH TOP FIRE TOWER

This tower is on Allegheny Mountain on the Virginia side of the ridge. DeGarmo (1953: 42) suggested that goodly numbers of hawks may appear in the area, but more observations are needed to evaluate the value of this spot.

MEADOW KNOB

Aside from Carpenter's (1960) statement that Meadow Knob, on Shenandoah Mountain, is used by hawk watchers, I can find no published information on this fire tower. The combined hawk counts for Shenandoah Mountain (table 41) contain some data from this lookout, however.

MENDOTA FIRE TOWER

This fire tower is in the extreme southwestern corner of Virginia near the Tennessee border. It has been used as a lookout by hawk watchers for many years. Yearly counts have appeared in *The Migrant* as part of the Tennessee Ornithological Society's autumn hawk counting project. Thus it seems best to consider the Mendota Fire Tower as part of the Tennessee project. Refer to the Tennessee section.

PURGATORY MOUNTAIN OVERLOOK

This lookout, elevation 2,415 feet, is on the Blue Ridge Parkway near milepost 92. Excellent hawk flights were seen here in mid-September 1972 (Beck, 1972), suggesting that future use of this spot by hawk watchers might be worthwhile.

REDDISH KNOB

Reddish Knob, elevation 4,398 feet, is in Augusta County. Use of the 50-foot fire tower as a hawk lookout first occurred on 17 September 1949 (Carpenter, 1949), and the site has been used during recent years (Carpenter, 1960). During the period 1949 through 1959, Carpenter (1960) reports combined yearly hawk counts ranging from 29 birds in 1958 to 715 birds in 1957.

WEST VIRGINIA

There has been considerable interest in hawk migrations in West Virginia since 1947 with particular emphasis on Broad-winged Hawks. Little information has been published, however. DeGarmo (1953: 40) states that "There is undoubtedly some seasonal movement of hawks throughout West Virginia. The writer has, for example, seen Red-tailed, Sharp-shinned, Broad-winged and Sparrow Hawks moving over the broken hills of Brooke County. There are records of sizeable flocks of Broad-wings in Upshur, Nicholas, and Kanawha Counties. Migration of this type is widely dispersed. It is not until long mountain ridges are encountered that great concentrations can be expected in West Virginia. Under certain weather conditions there is probably not a mountain ridge, extending in a north-south direction, which does not receive a few migrating hawks. There are, on the other hand, a number which are used more or less regularly as migration routes by hundreds of hawks annually. These are the ridges which hawk watchers have been attempting to find, so as to be able to reduce chances for a poor count during future years of migration studies." DeGarmo (1953: 42–43) also states that dates of West Virginia's hawk migrations generally correspond to the seasonal distribution of flights at Hawk Mountain, Pennsylvania.

Reports detailing West Virginia's hawk migrations are those of DeGarmo (1953), Hurley (1970), and Shreve (1970). Unpublished counts from various lookouts also are filed at Patuxent Wildlife Research Center.

ALLEGHENY FRONT

The importance of the Allegheny Front to migrating birds is well known. Hall (1964) describes the area as ". . . a mountainous escarpment of variable height, arising in central Pennsylvania, and continuing south into West Virginia, interrupted only by the Potomac River. It forms the boundary between the highly folded mountains of the Ridge and Valley Province to the East and the mountainous region known as the Appalachian or Allegheny Plateau to the West. As a major physiographic boundary it forms a rather natural flight line for such bird migrations as typically follow such barriers. The prominence of the ridge as a hawk flyway arises from this fact. Typically the mountain consists of a steep eastern face overlooking a valley some 1,500–2,000 feet lower than the ridge, a broad nearly flat top of some extent, and a more gentle descent to the West." Bear Rocks is one of the important hawk lookouts on the Allegheny Front.

BACKBONE MOUNTAIN

This mountain is in Tucker County in northcentral West Virginia. DeGarmo (1953: 41) states that it is the western edge of the Alleghenies and continues southward from Big Savage Mountain in Pennsylvania. He considers it "on a par with the Blue Ridge, North Mt., and Allegheny Front" but provides no hawk counts.

BEAR ROCKS

Bear Rocks, a rocky outcropping on the Allegheny Front, is a favorite lookout for hawk watchers (DeGarmo, 1953: 41; Hall, 1964). Very little hawk count information has been published for this site, however. Because it has little western slope, poor counts are recorded with strong prevailing west, northwest, or southwest winds—a condition opposite to that at Hawk Mountain, Pennsylvania—but when moderate winds occur from these directions, hawks readily use developing thermals. Bear Rocks is equally productive when moderate easterly and southeasterly winds occur.

Bear Rocks is located just north of the border of the Dolly Sods Scenic Area in Monongahela National Forest. The Red Creek Camp Ground off Forest Service Road No. 75 a few miles south of Bear Rocks can be used by hawk watchers.

Bear Rocks, W. Va., with the Dolly Sods area in the background. Photo courtesy of the U.S. Forest Service.

BLUE RIDGE

This is the easternmost mountain ridge in West Virginia and is one of the best for watching hawk migrations (DeGarmo, 1953: 40). He states: "It appears to collect birds from the Catoctin and South Mountain ranges in Maryland and possibly from Sterrat's [sic] Gap in Pennsylvania. In both 1951 and 1952 it produced good flights. Once this flight reaches the Blue Ridge it appears to continue southward a considerable distance, as shown by the high count obtained by Scott, Stevens, and Watson on Big Flat Mt. near Afton, Virginia on September 21, 1952." DeGarmo suggested that the Blue Ridge consistently produces good hawk flights. This conclusion was based on limited observations, however.

CHEAT MOUNTAIN

DeGarmo (1953: 41) considered Cheat Mountain an important hawk migration route in the central Alleghenies but mentioned that good observation spots were difficult to reach. However, Mace, near the southern end, was mentioned as producing a record count in September 1951.

HANGING ROCKS FIRE TOWER

Since 1952, the Hanging Rocks Fire Tower has proved to be one of West Virginia's important hawk lookouts (Hurley, 1970). The tower is located on Peters Mountain, elevation 3,812 feet, about three miles northwest of Waiteville. On the northwest side the mountain drops for 1,300 feet to the valley floor whereas the southeastern side gradually drops toward Waiteville at an elevation of 2,200 feet. Hurley's (1970) yearly counts of migrating hawks are summarized in table 43.

KNOBLEY MOUNTAIN

This site is located in Mineral County. According to DeGarmo (1953: 41), a flight of 503 hawks was observed here on 17 September 1952, but additional observations from Knobley Mountain proved unsatisfactory.

MIDDLE RIDGE

Middle Ridge is an open hilltop west of Kanawha State Park, Kanawha County. At an elevation of 1,160 feet, it is one of the highest spots in the county. Hawk watching was begun in 1969 by Shreve (1970). Three days of observation produced the following: 4 Turkey Vultures, 3 Cooper's Hawks, 5 Red-shouldered Hawks, 3,548 Broad-winged Hawks, 1 Osprey, 10 Sparrow Hawks, and 3 unidentified hawks.

NORTH MOUNTAIN

This is a high mountain range in Berkeley County rising abruptly south of the Potomac River and extending the length of the county, then ending suddenly. DeGarmo (1953: 40) reported 1,030 hawks from the fire tower on the mountain on 17 September

1950, but later counts were not as large. However, observations suggest that considerable movements of hawks occur along the valleys on either side of North Mountain. DeGarmo mentions that Lloyd Poland found good numbers of Broad-winged Hawks along North Mountain prior to 1950, suggesting that it might be one of the better West Virginia ridges from which to count migrating hawks.

NORTH RIVER MOUNTAIN

See the Hanging Rocks Fire Tower section.

PETERS MOUNTAIN

A detailed summary of autumn hawk flights along Peters Mountain is presented by Hurley (1970). Also refer to the section on the Hanging Rocks Fire Tower.

SLEEPY CREEK MOUNTAIN

According to DeGarmo (1953: 40), fair numbers of hawks have been counted here. Little else is known about the spot.

7

Southern Appalachian Hawk Migrations

Portions of the southern Appalachian Mountains are important as a terminal point and winter range for some migrating raptors, e.g., many eastern North American Golden Eagles, and an essential link in the migration routes of other species, notably Broad-winged Hawks. Considerable hawk watching has been done in eastern Tennessee in particular where, according to Thomas W. Finucane (letter of 5 May 1973), most of the hawk migrations in the state occur.

KENTUCKY

Although hawks migrate across Kentucky during autumn, Mengel (1965: 200–223) suggests that the spectacular flights which occur elsewhere in the eastern United States may not occur here. More study from fire towers and mountain lookouts is required to confirm this, however.

TENNESSEE

The spectacular flights of Broad-winged Hawks in mid-September long have been an exciting part of the birding year in Tennessee.

Hawk watches are coordinated by the Tennessee Ornithological Society. The initial leadership was by Fred W. Behrend from 1951 through 1954, and by Thomas W. Finucane during the years since 1955. Annual reports of these statewide counts appear in *The Migrant* (Behrend, 1951, 1952, 1953, 1954a, 1954b; Finucane, 1956, 1957, 1958, 1959, 1960, 1961a, 1961b, 1963, 1964, 1965, 1966, 1967, 1968, 1969, 1970, 1971, 1972).

Several important differences separate Tennessee hawk counts from those made elsewhere in the eastern United States. In Tennessee, virtually all studies are restricted to September and very early October, and primarily to Broad-winged Hawks. These flights are scattered over broad migratory fronts. Seldom does any single lookout consistently produce large hawk counts year after year although some lookouts are used regularly and produce fairly good flights. Moreover, a large percentage of Tennessee's hawk counts is made from fire towers scattered throughout the eastern half of the state. Finally, these counts are combined and published as a composite of all individual Broad-wing totals for Tennessee. Hence it is difficult to compare hawk counts from this state with those from other areas without separating or extracting the information for each lookout from the published summary. Table 44 summarizes Tennessee Broad-winged Hawk counts from 1951 through 1971 based upon the reports published by Behrend (1951 to 1954b), Finucane (1956 to 1972), and Odom (1966). When minor differences in yearly totals occurred between the latter author's information and the primary reports of the former authors, the data published by the former authors were used.

In addition to the references previously mentioned, a few other accounts of hawk migrations in Tennessee have appeared. Behrend (1950a) published a short account of Broad-wing flights over Hump Mountain southeast of Elizabethtown, and Herndon (1949) reported a Golden Eagle from this same site. Behrend (1950b) also provided more information on hawk flights in upper east Tennessee. Dunbar (1950) also reported on a Knox County hawk flight, and Ganier (1951) observed 130 Turkey Vultures migrating near Ashland City west of Nashville. Finally, Johnson (1950) published a note on hawk migrations along the Blue Ridge, and Spofford (1949)

reported on a Broad-wing flight at Fall Creek Falls State Park near
Pikeville.

Through the cooperation of Thomas W. Finucane (personal
communications), a few comments regarding some of the better
sites are possible although detailed descriptions of these lookouts
are unavailable.

DUNLAP FIRE TOWER

This tower is located on Walden Ridge north of Chattanooga
where U.S. Route 127 crosses the time-zone boundary. It is unusually accessible since it is by the side of the road.

ELDER MOUNTAIN

Until it was torn down and replaced by residential construc-
tion, the Elder Mountain Fire Tower (elevation 1,880 feet) just west
of Chattanooga was one of Tennessee's best lookouts. Excellent
Broad-winged Hawk flights have been seen from this site.

FALL CREEK FALLS STATE PARK FIRE TOWER

The fire tower in Fall Creek Falls State Park, on the Cumberland Plateau on Route 30 near Spencer, provides easy access to a
good lookout for hawk watchers. Spofford (1949) reported Broad-
winged Hawk migrations from this site.

MENDOTA FIRE TOWER

Although this tower is in Virginia near the Tennessee border,
it is used annually by the Tennessee Ornithological Society as a
lookout, and counts from Mendota are included in the statewide
Tennessee hawk counts. Thomas W. Finucane (personal communication) considers this site the principal lookout in the Tennessee
hawk watching project.

"The Mendota Fire Tower may be reached by driving north
from Abingdon, Virginia on Route 19 (Alt. 58) to Hansonville. At
Hansonville follow Route 802 to Route 614 and continue to the top

The Mendota Fire Tower, Va. This site is near the Tennessee border. It is used yearly by the Tennessee Ornithological Society in its statewide autumn hawk count. Photo by Tommy Swindell.

of the mountain. Park on top, in the saddle, and hike along a trail on the right to the fire tower. The hike requires about 15 minutes" (Heintzelman, 1972d: 60).

ROGERSVILLE-KYLES FORD FIRE TOWER

This fire tower is located north of Rogersville near Edison off Route 70. A ten-minute hike is required from the road to the tower, which is a good hawk lookout.

8

Southern Coastal and Gulf Coast Hawk Migrations

Ornithologists have a fragmentary knowledge of hawk migrations passing through the southern coastal and Gulf coast states. In the early 1950's Van Eseltine (1952) and Simpson (1952) commented on hawk watching in South Carolina and North Carolina, respectively. Later Simpson (1954) published a general account of the status of migratory hawks in the Carolinas, and a few other short notes concerning hawk watching have appeared in *The Chat*. Wray and Davis (1959) also provide cursory comments regarding North Carolina hawk migrations as do Sprunt and Chamberlain (1970) for South Carolina. Few reports are published regarding hawk flights in Georgia (Burleigh, 1958) and Florida (Howell, 1932; Sprunt, 1954), with the exception of a few references in *Audubon Field Notes*, and very limited information is available on the Gulf coast region.

NORTH CAROLINA

Apparently hawk migrations occur inland along North Carolina's mountain ridges as well as along the outer banks. Accipitrine and soaring hawks tend to follow the mountain routes, whereas falcons and some accipiters appear along the coast.

BLUE RIDGE PARKWAY

According to Simpson (1954: 17–18), the first report of a massed flight of Broad-winged Hawks in North Carolina was made by W. M. Johnson, who counted 2,322 birds on 25 September 1950 between Little Switzerland, Mount Mitchell, and Thunder Hill over the Blue Ridge Parkway. In 1952, a hawk watching project was begun in North Carolina with various lookouts in the Parkway being used, including Blowing Rock, Doughton Park, and Roaring Gap. Simpson (1952) reported small flocks of hawks being seen. Later Simpson (1954: 18) provided a few more Broad-wing counts from a letter which C. W. Senne of the National Park Service sent to Maurice Broun. Detailed hawk counts in the Blue Ridge Parkway are unavailable, however.

OUTER BANKS

Since the early 1950's, the Outer Banks of North Carolina have been known as a flyway for migrating Peregrine Falcons during September and October (Simpson, 1954: 20–21). Flights of Sharp-shinned and Cooper's Hawks also have been observed in the vicinity of Nags Head. Doubtless these offshore and coastal hawk movements contain some birds from the flights reported at Assateague Island, Maryland (Crossan and Stevenson, 1956).

TABLE ROCK

According to Pratt (1967), Table Rock forms the summit of the mountain located on the eastern edge of the Blue Ridge. It is located in northwestern Burke County. On 16 September 1967, hawk watchers counted 251 Broad-wings within a few hours. Simpson (1952) previously reported hawk flights from Table Rock as have various other observers who visited the spot occasionally.

TOPSAIL ISLAND

This island is located on the North Carolina coast near New River Inlet. On 8 October 1967, Gilbert S. Grant (1967) reported

29 Ospreys migrating over the island. This count was based upon a couple of hours of observation during the late morning and early afternoon. The birds were flying in a southerly direction, using alternating flapping and sailing flight. A northwest wind at 10 to 15 knots prevailed at the time. Also exhibiting migratory behavior were two Sharp-shinned Hawks and a Marsh Hawk.

SOUTH CAROLINA

Although Sprunt and Chamberlain (1970) provide general comments on South Carolina hawk migrations, little detailed information is published. The first organized hawk count occurred in 1951 (Van Eseltine, 1952), but the results were mostly negative. Parks (1957) provided fragmentary data for the 1957 season.

GEORGIA

Some autumn hawk migrations unquestionably cross Georgia, but these movements are scattered and difficult to detect judging from Burleigh's (1958) general distributional accounts. If systematic efforts were made in the northern mountains, and perhaps along the coast, some hawk flights might be seen, however.

FLORIDA

Although considerable numbers of hawks migrate into Florida and winter there (Howell, 1932; Sprunt, 1954), these hawk migrations generally were viewed as being widespread rather than concentrated. Within recent years some concentrations of hawks have been seen, however. This is particularly true of kettles of Broad-winged Hawks over Key West and other sections of southern Florida. Sharp-shinned Hawks and other species also have been reported during some autumns in southern Florida. Robertson (1970, 1971, 1972) and Robertson and Ogden (1968) commented in detail upon these remarkable (and unexpected) flights. There are two particularly significant aspects of the autumn Broad-winged Hawk movements observed in southern Florida and the Key West area. First, some of these birds (along with Sharp-shins and other diurnal raptors) also have been seen during mid- to late autumn by observers stationed on the Dry Tortugas (Robertson, 1972). Secondly, Tabb (1973: 11) reports that "substantial numbers" (census data not provided) of Broad-wings winter in southern Florida between Miami and Key West. Although Tabb (1973: 16–17) suggests that Broad-wings "commonly" migrate over the open waters of the Straits of Florida, in fact only limited detailed information is available on these flights. Indeed the appearance of Broad-winged Hawks over the Dry Tortugas is of considerable interest as this species seems almost never to cross large stretches of open water elsewhere.

Much additional field study is necessary before these Broad-wing flights can be understood properly. Radar tracking of the birds leaving Key West might be of exceptional value in determining the ultimate fate of the hawks as they fly over open water.

ALABAMA

Imhof (1962: 37, 182–185) provides some information on Broad-wing flights. He states that hawks generally follow mountain ridges but some appear at Fort Morgan and on Dauphin Island. His map shows a general southwestward hawk movement across the northern half of the state. A record of a flight of 473 hawks of 11 species on 10 October 1957 in the Mobile Bay Area also is filed at Patuxent Wildlife Research Center.

LOUISIANA

Little is known about hawk flights in Louisiana. Lowery (1960) provides some information, and later commented that large concentrations of Broad-wings sometimes occur in extreme southern Louisiana (George H. Lowery, Jr., letter of 3 January 1969). Some falcons may also engage in a trans-Gulf migration, but only one Broad-wing record (spring) is known over the waters of the Gulf of Mexico (Lowery and Newman, 1954: 536). McIlhenny (1939) reported a flight of less than 500 Broad-winged Hawks over Avery Island on 22 September 1938. Oberholser (1938) provided general comments on hawk migrations in Louisiana.

TEXAS

The migration of Broad-winged Hawks through Texas during autumn is known in general terms. Van Tyne and Sutton (1937) did not include this species on their list of birds for Brewster County, but Wauer (1973) has recorded the species in the Big Bend National Park area on four occasions in autumn. Wolfe (1956: 19) states that Broad-wings are migrants ". . . in the central and southern part of the state and south to Cameron County." Peterson (1960) further states that these hawks migrate through the eastern half of Texas, rarely reaching west to the Panhandle and south to the Staked Plains.

Eastern Texas, therefore, serves as the mouth of the funnel which directs southward-bound Broad-winged Hawks into Mexico and Central America. Some individuals remain in southern Central America, whereas others continue into northern South America, which also serves as part of the winter range for this species.

9

Central American
Hawk Migrations

Some hawks which migrate southward across eastern North America winter in Central and/or South America. In some instances, species which form part of the eastern North American autumn hawk flights also occur as migrants or residents in Central America, but are represented there by subspecies other than those forming the North American migratory population. The Turkey Vulture is an example. Wetmore (1965: 161–167) lists three subspecies, one resident and two migratory, for Panama: *C. a. aura, C. a. meridionalis,* and *C. a. ruficollis.* However, *C. a. septentrionalis,* which forms the migratory eastern North American population, is not found in Panama. Similar examples could be cited for other raptors. Therefore, this chapter is restricted mainly to the autumn migrations of the nominate race *
of the Broad-winged Hawk which pass through Central America.

Although hawk migrations have received limited attention in Central America, available information indicates that this land bridge between two great continents is the migration route used by Broad-winged Hawks flying to their winter range, which is largely

---

* In a polytypic species, individuals of the first subspecies described sometimes are referred to as members of the nominate race.

in northern South America and southern Central America. Since Broad-wings are diverted around large bodies of water, upon reaching the Gulf Coast they are funneled through Texas into Mexico and southward through Central America. Much information still is required on these migrations, however.

MEXICO

Friedmann, Griscom, and Moore (1950: 55–56) record the nominate race of the Broad-winged Hawk in southern and southwestern Mexico only. However, large flocks of migrating Swainson's Hawks occur in Veracruz. Since mixed flocks of Broad-winged and Swainson's Hawks commonly occur elsewhere in Central America, some of the Veracruz flocks may contain some Broad-wings. Edwards (1972) considers the Broad-wing an irregularly common transient in Mexico, and Peterson and Chalif (1973) state that large loose flocks migrate through the northeastern and southern sections of the country except the Yucatan Peninsula. A few birds also may winter in Mexico.

BELIZE (BRITISH HONDURAS)

The only Broad-winged Hawk record for this country is a female of the nominate race collected by Peck on 22 October 1906 at Toledo Settlement in the extreme southeastern portion of the country (Russell, 1964: 48).

GUATEMALA

Smithe (1966: 31) summarizes the status of the Broad-winged Hawk at Tikal and the Peten as an uncommon but probably regular visitant during the migration season. It occurred at Tikal on 8 October, 4 December, and 28 March. The autumn birds were in immature plumage and tended to be solitary. No mass migrations such as occur in Costa Rica were seen, however. Land (1970) lists the Broad-winged Hawk as an autumn and spring transient.

HONDURAS

According to Monroe (1968: 76), huge flocks of the nominate race of the Broad-winged Hawk migrate through Honduras "at least on the Pacific slope." He states that Broad-wings were the most abundant migratory hawks on the slope during the autumn of 1962. About 40 birds were counted on 3 October, and increasingly large numbers were noted on days following until a peak flight, estimated in excess of 6,000 birds, occurred on 6 October. The hawks travelled west to east, roughly parallel to the coast, and funneled into the Choluteca Valley just east of Choluteca. Typically, the birds rode thermals until they reached an altitude from which they could glide over the sides of nearby mountains 1,000 meters in elevation. They then passed to the Caribbean drainage near San Francisco and continued eastward into Nicaragua. Monroe noted that Turkey Vultures and Swainson's Hawks often were mixed with the Broad-wings.

EL SALVADOR

Limited information is available on Broad-winged Hawk migrations in El Salvador. Dickey and Van Rossem (1938: 111) state that this species is a common fall and spring migrant. The first birds they noted were scattered individuals among a flight of Turkey Vultures, Swainson's and Red-tailed Hawks over Divisadero on 12 October 1925. Broad-wings occurred in other hawk flights until 6 November, and other individuals wintered on Mount Cacaguatique.

NICARAGUA

No information is available. See comments by Monroe (1968) on hawk flights in the Caribbean drainage area of Honduras near San Francisco.

COSTA RICA

Slud (1964: 60–61) and Skutch (1945) report that the Broad-winged Hawk is seen in large flocks during migration. It winters in the country from October to April and is one of the few abundant hawks seen.

PANAMA

Ornithologists have given more attention to autumn hawk migrations crossing Panama than elsewhere in Central America. Eisenmann (1963: 245–246) suggested that Black Vultures (non-North American birds) may migrate through Panama, and pointed out that Broad-winged and Swainson's Hawks are prone to use mountain updrafts, thus accounting for flights occurring more inland rather than along the Pacific slope. However, ridges approach the sea on the Caribbean slope and "vast" October flights occur there. Wetmore (1965) provides general comments on hawk migrations.

During 1963 to 1965, at a bird banding station near Almirante, Bocas del Toro, on the Caribbean coast, Hicks, Rogers, and Child (1966) noted many migrant hawks:

October 7, 1963	More than 100 Broadwings overhead at 10 A.M.
October 8, 1963	Several hundred Broadwings high overhead at 9:30 A.M.
October 9, 1963	Broadwings roosting in cacao groves. One caught in mist net this morning.
October 10, 1963	Broadwings roosting around lab. Rising from woods 8:00–8:30 A.M.
October 11, 1963	Broadwings passed overhead for three hours. Impossible to estimate numbers, but at least several thousand.
October 14, 1963	Mixed flocks of Broad-wings, Swainson's, and unidentified kites roosting around lab early this morning.
October 7, 1964	Rainy and cool. No hawks.
October 8, 1964	Netted first Broadwing of season today. Strip of sky from NW to SE filled with migrating hawks, mostly Broadwings with a few Swainson's. Continued for 3 hours—cannot estimate numbers.
October 9–11, 1964	Cloudy and rainy. No hawks.
October 12, 1964	Hawks overhead in huge circling flocks, mostly Broadwings.
October 13–16, 1964	Rainy mornings, no hawks.
October 17, 1964	50–60 Broadwings at 10 A.M. in small groups of 7–10.

October 7, 1965   More than 50 Broadwings leaving trees and riding convection currents, 6 A.M., 20 miles SE of Almirante.

October 12, 1965   Three groups of Broadwinged Hawks swirling in updrafts at 9:30 A.M., each group consisting of 25–50 individuals. Hawks coming into updraft from all directions.

October 13, 1965   9:00 A.M.—Three groups in separate updrafts, coming in from all directions. 11:00 A.M.— Several groups of hundreds passing to the SE.

October 14, 1965   9:30 A.M. Large swirl of 50–75 Broadwings. 3–4 P.M.—Observed thousands of hawks very high passing to the SE.

October 15, 1965   9–10 A.M.—A few small swirls consisting of less than 50 Broadwings. 3:30–5:00 P.M.—Observed large strings of hawks migrating to the SE, estimated numbers greater than 1000. Also about 1000 Kites very high mixed with a few unidentified Buteos. Kites are either Mississippi (*Ictinia mississippiensis*) or Plumbeous (*I. plumbea*).

October 17, 1965   Observed thousands of migrating Black and Turkey Vultures with a few Buteos mixed in, all migrating across Almirante Bay to the SE.

Kleen (1969: 172) also witnessed impressive hawk migrations in Panama. "One of the outstanding sights while at Almirante was the hawk and vulture migration. I have never been to Hawk Mountain, Pennsylvania, but I know that this migration along the Caribbean coast is just as spectacular if not more so. Actual counts of over 100,000 Turkey Vultures (Oct. 28, 1967) would soar by endlessly on the rising thermals and then change direction right over the city of Almirante. During a two week period (Oct. 24–Nov. 7) the Black Vultures, Swainson's Hawks and Broad-winged Hawks would also drift past by the 1000's. Other hawks were noted, but not in such large numbers. It is encouraging to know that the Swainson's Hawks are still plentiful knowing how much they are persecuted in their summer range." Kleen (letter of 13 August 1969) indicates that he never observed hawks and vultures flying over the

A spectacular kettle of migrating Broad-winged Hawks over Ancon Hill, Canal Zone, on 11 October 1972. Photo by Neil G. Smith.

water. Rather, they were following the coast in a line which was nearly parallel to the railroad which runs from Guabito to Almirante. The birds were using thermals and rising over Almirante, then turning landward to a 90-degree angle from the direction in which they arrived. Without the change of direction, the buteos would have had to fly over the water (which Ospreys did).

During the autumn of 1968, Charles Leck (letter of 18 November 1968) noted very large hawk migrations passing over Barro Colorado Island in the Canal Zone:

October 9, 1968	Two Broad-winged Hawks observed.
October 10, 1968	About 0915 I started noting hawks going overhead; hawks passing close enough to be identified were Broad-winged Hawks. Weather—clear with scattered cumulus clouds; 83 degrees F. General movement of the birds was from the SW towards the NNE. There was *no wind*.

From 0915 to 1015, approx. 3,960 birds went overhead; and another 153 came before 1030, to bring the total to *4,113*. Almost none seen during the remainder of the day. They traveled in long lines (chains) . . . and would come in sizeable groups, sometimes forming kettles, and then drifting off again in a long chain. The groups . . . (in order, except for single birds) 25, 100, 470, 200, 300, 100, 25, 250, 280, 150, 170, 19, 200, 260, 260, 125, 200, 250, 100, 300, 400, 36, 97, 50, and six individuals.

Vultures were also noted in migration. Hundreds of Turkey Vultures were noted moving with the above hawks. At one time, 150 Turkey Vultures were visible. In general, they appeared scattered with the hawks, although the Vultures would sometimes kettle by themselves, lower than the hawk kettles. (In addition, local Black Vultures were about, and three —together—King Vultures were noted.)

October 11, 1968   Hawk migration again this morning (details much as above, with 2,984 estimated from 0845–1030). Almost all were seen between 0930–1030. Again, many Turkey Vultures in migration.

Another kettle of Broad-winged Hawks over Ancon Hill, Canal Zone. Photo by Neil G. Smith.

Broad-winged Hawks gliding from a thermal over Ancon Hill, Canal Zone. Photo by Neil G. Smith.

October 12, 1968	One Broad-winged Hawk noted.
October 13, 1968	265 Broad-winged Hawks were noted.
October 15, 1968	Ten Broad-winged Hawks were noted.
October 18, 1968	12 Broad-winged Hawks were noted.

Leck noticed that other migrant birds of prey were limited. He observed 3 Sparrow Hawks, 1 Peregrine Falcon, and a few Ospreys. However, on 2 November 1968 near Panama City, Panama, he saw a continual line of at least 1,500 migrating Turkey Vultures.

Smith (1973) also observed large migrations of Swainson's and Broad-winged Hawks over Ancon Hill in the Canal Zone during October of 1972, and published some photographs of several large kettles of Broad-wings. He reports counting 41,333 Broad-wings from Ancon Hill on 11 October 1972.

Part 3
Raptor Morphology, Anatomy, and Flight

10

Raptor Morphology and Anatomy

An understanding of autumn hawk migrations depends upon detailed observations of migrating birds as well as upon an examination of the morphology and anatomy of the various hawk species. That is, it is important to relate morphology and anatomy to function (migration). As used here, *morphology* refers to a bird's external shape and form, *anatomy* to internal structure as revealed by dissection.

In the past relatively little effort has been made to combine morphological and anatomical discoveries with the results of ornithological field studies (or vice versa). Among Falconiform birds, exceptions are found in the works of Chandler (1914), Engels (1941), Fisher (1946), Hartman (1961), and R. W. Storer (1955). Many topics still require careful investigation, however.

SHAPES OF RAPTOR WINGS

Of the eight genera and fifteen species of diurnal birds of prey with which I am concerned, each genus shows important differences in wing morphology and, on an intrageneric level, lesser differences. For example, our three *Accipiter* species have similarly shaped wings, but differences in their proportions usually enable an experienced hawk watcher to identify individual birds correctly on the basis of shape and wingbeat. Because variations in wing shape affect hawk migrations, the shapes are briefly examined.

GENUS *Cathartes*

Of the three species in this genus (Brown and Amadon, 1968), the Turkey Vulture is of concern here. These are large birds with wingspreads of about six feet. Their wings are long, moderately wide, and have strongly slotted tips. Typically the wings are held slightly above a horizontal plane when the bird is aloft. This forms a characteristic dihedral which is very useful in making correct field identification. When engaged in soaring and gliding flight, Turkey Vultures commonly rock and tilt unsteadily (Peterson, 1947).

Although Turkey Vultures use thermals, they are more dependent upon updrafts when migrating along mountains. The birds use the air currents skillfully and seldom exert much energy by flapping their wings.

A soaring Turkey Vulture showing the dihedral and the strongly slotted tips of its wings. Photo by Donald S. Heintzelman.

An immature Goshawk showing the slotting of the tips of its wings. Photo by Fred Tilly.

GENUS *Accipiter*

This genus contains forty-seven species (Brown and Amadon, 1968). I am concerned with the Goshawk, Sharp-shinned Hawk, and Cooper's Hawk.

Accipiters are woodland or forest dwelling hawks. Their wings are short and rounded and are fairly strongly slotted. In addition, accipitrine tails are long and rudderlike. The typical style of flight consists of a few rapid wingbeats followed by a period of sailing, then more rapid wingbeats. Each of the accipiters have similar flight patterns, but an experienced observer often can identify each species by observing slight differences in wingbeat and style of flight. There is usually no difficulty in identifying a hawk as an *Accipiter*, however. The Sharp-shinned Hawk, the smallest North American accipitrine species, has the most rapid wingbeat whereas the Goshawk, the largest, has the slowest wingbeat.

GENUS *Buteo*

This genus contains twenty-five species (Brown and Amadon, 1968). Considered here are the Red-tailed, Red-shouldered, Broad-winged, and Rough-legged Hawks. The Swainson's Hawk only oc-

A Red-tailed Hawk showing the slotting of its wing tips. Photo by Alan Brady.

casionally occurs in the northeast (e.g., see Graham, 1972). Eastern buteos are moderate to fairly large birds with broad, very strongly slotted wing tips and a broad, rounded tail which is spread when circling. Each species spends some time gliding and soaring during migration. In general, all but the Broad-winged Hawk prefer soaring on mountain updrafts. In contrast, the Broad-wing is dependent upon thermals but readily uses updrafts when engaged in inter-thermal glides. It also uses combinations of thermals and updrafts if the latter are not too strong. This species virtually is unable to fly in very gusty wind, however.

As in the genus *Accipiter,* differences between the eastern buteos usually allow experienced observers to identify individual birds correctly on the basis of flight style and body proportions. Patterns of plumage and coloration are equally important field marks.

GENUS *Aquila*

This genus contains nine species of eagles (Brown and Amadon, 1968). Only the Golden Eagle is of interest here. It is a large bird

A Golden Eagle showing the slotting of the tips of its wings. Photo by Steve Piper.

whose wings are held flat when in flight and which are proportionally longer than buteo wings. *Aquila* wing tips are very strongly slotted. My impression of the flight of Golden Eagles is that they do much soaring, but when engaged in flapping flight the stroke is powerful, slow, and almost labored.

GENUS *Haliaeetus*

This genus contains eight eagles (Brown and Amadon, 1968). I am concerned with the Bald Eagle. They have very flat wings which are often fully extended when maximum lift is needed, but at other times the primaries are commonly "pulled back" as a bird soars or sails on updrafts. As in *Aquila,* the wing tips are very strongly slotted. The flight patterns of the two eagles also are similar. I am unable to distinguish any obvious differences in the overall proportions of the wings of the two eagles. Similarly, Broun (1949a: 190–191) is unable to notice any obvious differences in the wingbeat, speed, or manner in which Golden Eagles and Bald Eagles hold their wings. Many experienced observers have commented on the differences in the size of the heads and bills of the two eagles, however. Bald Eagles have much larger heads and bills than Golden Eagles.

GENUS *Circus*

This genus contains ten species (Brown and Amadon, 1968). The Marsh Hawk is of interest. These harriers have long wings

A Marsh Hawk showing the slotting of its wing tips. Photo by Donald S. Heintzelman.

which appear somewhat pointed and not too unlike those of falcons. Indeed, inexperienced observers seeing Marsh Hawks flying at a distance must be careful not to identify the birds incorrectly as falcons. However, the tips of the wings of Marsh Hawks are moderately slotted, and these birds have a characteristic flight pattern involving considerable tipping and tilting, which quickly enables one to separate a harrier from a falcon. Marsh Hawks, in addition to having different plumage patterns from those of our falcons, also hold their wings in a slight dihedral.

A typical Marsh Hawk flight characteristic is the tendency to fly on an unsteady zigzag course along mountain ridges regardless of wind direction.

GENUS *Pandion*

The Osprey is the only species in this genus (Brown and Amadon, 1968). They are large, magestic birds with long, moderately wide, very strongly slotted wings. Ospreys sometimes circle with their wings held fully extended. Most commonly during migra-

An Osprey showing the slotted tips of its wings. Photo by Donald S. Heintzelman.

tion, however, they soar on updrafts with their wings bent into a deeply formed crooked position. This gives the birds an extremely characteristic appearance. When Ospreys are seen approaching at eyelevel, their long wings have a downward-bowed appearance which also is characteristic and very useful as a field mark.

GENUS *Falco*

This large falcon genus numbers thirty-seven species (Brown and Amadon, 1968). I am concerned only with the Peregrine Falcon, Pigeon Hawk, and Sparrow Hawk. Although the Gyrfalcon occasionally appears at some of the Great Lakes hawk lookouts, it rarely appears elsewhere and is not of important consideration in studies of autumn hawk migrations.

The outstanding characteristic of a falcon's wings is their long, pointed shape. Peterson (1947) states that these birds are built for speed and not for soaring but, when they do soar, their wings lose some of their pointed effect. Falcon wings are only moderately

A Sparrow Hawk showing the slight slotting of its wing tips. Photo by Donald S. Heintzelman.

slotted. Each of our eastern species has a rowing style of flight, most conspicuous in the larger species.

ASPECT-RATIO OF RAPTOR WINGS

The efficiency of a bird's wings depends upon a number of factors. To attain its greatest lift, for example, a wing must be streamlined and as long as possible to spread the tip vortices far apart, which provides as much undisturbed area between them as possible. In addition, a wing must be wide enough to stop the eddies along the posterior edge from seriously interfering with the efficiency of the forward lifting area. Thus a wing of an airplane, or of a bird, is a compromise between the necessity for the strength provided by a short wing and the need for the efficiency which is provided by a long wing (J. H. Storer, 1948). Although several basic wing forms have evolved in birds to permit successful exploitation of different ecological niches (Saville, 1957), the variations on these basic wing types are almost limitless.

One of the most convenient ways to describe a bird's wing is to refer to its aspect-ratio. This is the ratio of the length of the wing to its width (J. H. Storer, 1948: 9) calculated via the formula (Hartman, 1961),

$$\text{Aspect-Ratio} = \frac{\text{Length of Wing}}{\text{Width of Wing}}$$

The long narrow wings of an Osprey have a relatively high aspect-ratio. Photo by Donald S. Heintzelman.

where the width is meant to be the median width. Hartman (1961: 26) points out that the aspect-ratio of a bird's wing indicates its adaptation for soaring or quick takeoff. High aspect-ratio wings have evolved for soaring whereas low aspect-ratio wings are adapted for prompt takeoff.

Only fragmentary data are available on the aspect-ratios of bird wings, however. Hartman (1961) provides the most complete summary (table 45).

WING LOADING

Another factor affecting the efficiency of a hawk in flight is the bird's wing loading (Brown and Amadon, 1968: 56–62). According to Hartman (1961: 32) wing loading can be calculated from the formula,

$$\text{Wing Loading} = \frac{\text{Body Weight in Grams}}{\text{Wing Area in Square Centimeters}}$$

Brown and Amadon (1968: 59–62) found that certain generalizations apply to wing loading in birds of prey.

1. There appears to be a general increase in wing loading with an increase in weight.

2. Certain genera, e.g., *Buteo* and *Circus,* have similar wing loading despite variations in size.

3. Birds of prey with buoyant flight seem to have rather low wing loading.

4. Most birds of prey which undertake long-distance migrations have low wing loading, especially if they cross open water. Exceptions exist, however.

5. In general, soaring hawks which undertake long-distance migrations and have high or fairly high wing loading use thermals as aids in their migrations.

Wing loading, of course, does not control the migratory range for any species, but low wing loading may assist in allowing a bird to undertake long migrations.

The two primary sources of wing loading data are Poole (1938) and Brown and Amadon (1968: 56–62). Hartman (1961) also provides a good deal of related data, but it is impossible to recalculate wing loading from his information. Moreover, R. J. Clark (1971) pointed out correctly that Poole's (1938) figures really are not wing loading data in the generally accepted sense and that Poole himself referred to them not as wing loading data but rather as wing area per weight data. However wing loading, expressed as grams per square centimeter, easily can be recalculated from Poole's valuable data. This is done for the information presented in table 46. In this table, averages are presented when more than one bird was available.

Although wing loading is a frequently used measurement in aerodynamic studies, other methods of expressing avian wing areas and bird weights have been used. Poole (1938), J. H. Storer (1948: 78), and Hartman (1961), for example, use the ratio of the wing area per body weight. This is the technique to which R. J. Clark (1971) addressed his comments and recommendations. Poole's (1938) data are summarized for diurnal birds of prey in table 47.

TAIL AREA, GLIDE AREA, AND BUOYANCY INDEX

Hartman (1961) developed three additional descriptive expressions of the shapes and efficiencies of raptors: (1) the area of a bird's tail, (2) the glide area or combined areas of a bird's wings, tail, and body, and (3) the buoyancy index expressed via the formula,

$$\text{Buoyancy Index} = \frac{\sqrt[2]{\text{Wing Area}}}{\sqrt[3]{\text{Body Weight}}}$$

Table 48 presents Hartman's calculations of each of these expressions for various raptors. A low buoyancy index is best for birds requiring sustained fast flying, whereas a high index is desirable for birds which soar and glide frequently.

THE HEART

A bird's heart is of major importance because it determines the animal's activities (Hartman, 1961). Without adequate blood circulation, provided by the heart as a pump, a bird's muscles could not function. The size and weight of the hearts of various birds of prey thus may be of importance in providing insights into the styles and methods of flight used by raptors during migrations. For example, Hartman points out that "heart weight is directly related to the ability to sustain power flight." Small hearts limit activity whereas those which are larger can maintain it much longer.

Among the vultures, for example, the hearts of Florida's Black Vultures ($P<0.05$) and Turkey Vultures ($P<0.01$) are larger than those living in Panama (Hartman, 1961: 7). Although subspecific determination of Hartman's specimens was not made, Wetmore (1965: 161–167) demonstrated that three subspecies of the Turkey Vulture occur in Panama—one resident (*ruficollis*) and two migratory (*aura* and *meridionalis*). Presumably migratory birds would have larger hearts, as may be suggested by Hartman's comparison of Florida and Panama specimens, since migratory birds might engage in longer and more strenuous flight.

Unfortunately relatively little is known about the movements of Black Vultures. Although generally considered nonmigratory, one passed over Hawk Mountain, Pennsylvania, in mid-November of 1969 (Nagy, 1970: 6), and Eisenmann (1963) presented information suggesting that Black Vultures are migratory in November in Panama. Wetmore's (1965: 161) observations of Black Vultures in Panama do not verify the migratory behavior of this species, however. In any event *Coragyps atratus brasiliensis,* the subspecies recognized by Wetmore (1965) in Panama, does not occur in eastern North America. Nonetheless, Hartman (1961: 7, 42) demonstrated that differences do occur between body weights (which are heavier) and heart size (which are larger) among Black Vultures in Florida as opposed to birds living in Panama.

Data on heart size are not available for all of the species considered in this book, but adequate information is available to permit preliminary comparisons. Table 49 summarizes Hartman's (1961) data.

Curiously, Red-shouldered Hawks examined during June and again during winter in Florida exhibited striking seasonal variations in relative heart size (Hartman, 1961: 17–18, 88). Seven January and February birds had hearts whose percentage of body weight was 0.73 ± 0.05 whereas hearts from seven June specimens were 0.55 ± 0.03 ($P<0.01$). This suggests that the size differences between sedentary and migratory races could account for the different values obtained during different seasons of the year. On geographic grounds three subspecies of the Red-shouldered Hawk (*alleni, extimus,* and *lineatus*) occur in Florida at various seasons (American Ornithologists' Union, 1957). The subspecies *lineatus,* for example, is the largest of the races reaching Florida (Friedmann, 1950). It occurs there during winter, the period when Florida Red-shoulders showed their highest values in Hartman's study. Observers in Florida have noted that some autumn Red-shouldered Hawks are "much larger and darker than local birds" (Robertson, 1971: 46). These probably are representatives of the subspecies *lineatus.*

THE STERNUM

In considering the migrations of hawks, it is also desirable to examine the parts of the skeletal system which are most directly related to flight—the sternum and its keel—since these bones serve as anchor points for the major flight muscles. Indeed, George and Berger (1966: 10) state that the degree of relative development of the sternum's keel is directly related to the degree of development of two major flight muscles—Mm. pectoralis and supracoracoideus.

Among Falconiform birds R. W. Storer (1955) made detailed studies of sternum variations and differences in the Goshawk, Sharp-shinned Hawk, and Cooper's Hawk. He pointed out that objects of similar shape, surface area, and volume do not increase directly with linear proportions. Rather surface area increases as the square of linear proportions, whereas volume or weight increases as the cube of linear proportions. To test the concept, measurements were made and compiled for the three accipitrine hawks (table 50).

It is common knowledge that *Accipiter* females (and females of most other Falconiform genera) are larger than males. Thus Storer found that the three eastern North American hawks in his study form a series of size groups ranging in weight from 100 grams for males of *velox* (Sharp-shin) to 1,100 grams for females of *atricapillus* (Goshawk).

In respect to the relationship between wing area and weight, a line with an expected slope of two to three exists, the square (wing area) versus the cubic function (weight). Storer points out that the relationships between the Sharp-shin and the Cooper's Hawk, the two smaller species, are rather good but there is a slight disagreement in the case of the Goshawk because a single measurement of wing area was available. In the three hawks studied, wing loading increased from 0.24 for a male Sharp-shin to 0.57 for a female Goshawk.

MAJOR FLIGHT MUSCLES: PECTORALIS AND SUPRACORACOIDEUS

George and Berger (1966: 22) list some fifty different striated muscles and muscle slips which act on the bones and feathers of a bird's wing. The two which play the major roles in depressing and elevating the humerus, and thus the whole wing, are Mm. pectoralis and supracoracoideus. Because of the importance of these muscles to bird flight, it is desirable to tabulate their weight (table 51) expressed as a percentage of body weight (Hartman, 1961).

Hawks with strong powers of flight, and long-distance migrants such as the Peregrine Falcon, show very heavy pectoral muscles whereas soaring species such as the Broad-winged Hawk have light pectoral muscles. There are exceptions, however. Thus the Cooper's Hawk is a rather strong flyer, but its major flight muscles are light, very nearly approximating those of the Broad-wing. Curiously the Osprey, which is a strong flyer and a species which undertakes long migrations, often over open water, has major flight muscles which are only slightly heavier than those in soaring hawks. However, this species is capable of considerable soaring. This could be a related factor to the speculation (Heintzelman, 1970c) that pesticides or other factors may reduce the energy production of flight muscles in Ospreys, thus causing the birds to become more dependent upon updrafts at flyways such as Bake Oven Knob and Hawk Mountain, Pennsylvania. Hence higher counts of these birds have occurred at these locations in recent years.

MORPHOLOGY AND AERODYNAMICS

The aerodynamics of a flying bird is very complex. Cone (1962a) has explained the aerodynamics of a flying bird of prey in enough detail for nonengineers to understand and appreciate the basic mode of operation, however.

To begin, a bird in flight is subject to the same aerodynamic laws as is an airplane. When a wing increases its angle of attack, it increases its lift as well. Unfortunately its drag increases even more rapidly. Part of the drag, although small and unaffected by lift, is caused by the friction between the wing and the air. The other part of the drag is referred to as induced drag. It generally depends upon lift and is associated with variation of the lift force across the span of a wing. There is a decrease of lift from the maximum at the center of a wing to zero at the tip. This results in series of small vortices which are shed from the whole of the trailing edge of a wing, each vortex being proportional in strength to the rate at which the lift decreases at the point where the vortex is shed. As a bird passes through the air, vortices are continually generated with the kinetic energy per second put into the vortices equaling the power which is required for a bird to maintain flight. Cone (1962a) points out that the smaller the required power, the greater the ease with which a bird can soar. The higher the aspect-ratio of a wing, the weaker the vortices leaving the trailing edge. Thus lower induced drag results—a technique used in sailplanes, which have long and narrow wings. There is a practical limit to the use of aspect-ratio as a drag-reduction device, however.

In airplanes, another method of reducing wing-induced drag is the placement of endplates, or large discs, on the tips of the wings. These plates prevent lift from becoming zero at the wing tip and reduce the lift variation across the span so that induced drag is lessened.

To illustrate the principle in birds, Cone (1962a) compares two excellent soaring birds, a condor (species?) and an albatross (species?). The albatross has a very high aspect-ratio wing and is a superb soarer. The condor, on the other hand, also is an excellent soarer yet the aspect-ratio of its wing is *much less* than that of an albatross. It appears that the condor should hardly be able to soar, yet its low aspect-ratio wing has high aspect-ratio efficiency. Cone states that this is achieved via the endplate mechanism. A similar explanation applies to other diurnal birds of prey.

Slotted wing tips are common but unique structures on land-soaring birds with low aspect-ratio wings. Evidently slotted tips are intimately associated with induced drag. But how? Cone states that the condor's wing tip contains long, slender primaries which are essentially individual, high aspect-ratio air foils. When the condor is engaged in circling flight, the primaries are fully extended in a fan-shaped position. Large gaps or slots are thus created between each successive primary. In addition, the primaries also are spread out vertically as a result of their upward bend under the air load they carry. The leading primary is bent upward strongly, and its tip is nearly vertical. The next primary has slightly less curvature than the leading feather, and succeeding feathers are curved still less with the last primary actually curved downward when the bird is in flight. The first three condor primaries are all heavily loaded whereas the remaining are much less loaded.

When a bird is in flight the curvature of the primaries is automatic since they assume their shape as a result of their elastic deformation under the air loading—the bird does not have direct control over their curvature. Cone demonstrated this by testing a rigid vulture (species?) wing in a wind tunnel. "At angles of attack below stall and at speeds equal to the usual flight speed of the bird

Each of the Turkey Vulture's long narrow primaries is a high aspect-ratio air-
foil. Photo by Fred Tilly.

(20–25 miles per hour) the pinions automatically spread out vertically in *exactly* the same pattern as that photographed in soaring
birds. Different degrees of pinion spread and curvatures can be
obtained by varying the wind speed while holding the angle of
attack of the main wing constant or by varying the angle while
holding the air speed constant."

Cone (1962a) further points out that the angle of attack of a
bird in flight uniquely determines the equilibrium airspeed of the
bird, resulting in a natural combination which he refers to as "design" flight conditions. Each primary is a carefully tailored structure with a specific elasticity gradient and twist resulting in its
attaining a precise curvature for any specific angle of attack being
used. Interestingly, a primary not subjected to air loading curves
strongly downward at the tips, and the leading primary exhibits
an appreciable geometrical twist near its tip. Thus, it presents a
large angle of attack to the airstream which results in the sharp
curvature of the feather when it is engaged in heavy lift as the bird
is in flight. The next two condor primaries have decreasing amounts

of twist, and the remaining primaries have no twist. When a condor is in flight, the tips of its four outside (leading) primaries are bent from a strong downward curvature to an upward curvature. An appreciable force is required to do this, and for large birds such as the condor a pound of force may be required to bend a single primary. When lower lifts are involved, the curvature of the feather automatically decreases, and the wings assume more conventional tip shapes. When the primaries bend and curve under an air load, various stresses are set up in the vanes of the feathers which cause them to become increasingly cambered as they bend. Cambered airfoils produce much more lift than flat sections, thus explaining why the primaries become highly curved at low flight speeds. The ultimate in aerodynamic design results—the use of aeroelasticity for automatic geometry control.

Cone (1962a) points out that nature has apparently achieved the ultimate exploitation of the endplate principle in using slotted wing tips as a drag reduction device without encountering the disadvantages of high aspect-ratio wings or large endplates.

FLIGHT SPEEDS

Diurnal birds of prey exhibit a considerable variety of modifications on basic morphological and anatomical design. Hence it can be expected that these variations will be reflected in the flight styles and speeds of the different species. However, the flight speed measurements obtained at Hawk Mountain, Pennsylvania, during the autumn of 1942 do not support this hypothesis (Broun and Goodwin, 1943). Each of the records contained in table 52 is of a bird flying along a straight, uninterrupted course. The range and variation of ground speeds is a reflection of various wind conditions existing at the time the birds were timed. For example, Sharp-shinned Hawks rarely flapped when wind speeds increased above 15 miles per hour. The birds folded their wings against their bodies and were essentially using diving flight as they rode the updrafts. Similarly, an Osprey flying at 80 miles per hour was using a strong thermal and was involved in steep diving flight without a loss of altitude as it was being timed. Thus the high correlations which one might expect between different morphological variations and flight speed were lessened by local weather conditions prevailing at the time the flight speeds were made. Nonetheless, it is common knowledge that buteos are relatively slow-flying hawks, falcons are swift birds when aloft, and so forth.

Perhaps more important than speed is the relationship between morphology and anatomy and the migration routes used by diurnal birds of prey. In general, soaring hawks of the genus *Buteo* tend to use inland mountains, where good updrafts or thermals occur, rather than coastal areas. Moreover, these species usually avoid crossing large expanses of open water. However, strong-flying species such as Ospreys and falcons readily fly over open expanses of water. Since it would be desirable to cross bodies of water as rapidly as possible, flight speed is one factor influencing selection of migration routes used by various birds of prey.

Part 4
Hawk Migrations and Weather Conditions

11

Hawk Migrations and General Weather Systems

To understand autumn hawk migrations in eastern North America, one must examine the relationship of general weather systems to hawk movements. Allen and Peterson (1936), Broun (1935, 1939), Ferguson and Ferguson (1922), Poole (1934) and Trowbridge (1895), for example, early recognized the importance of north-westerly winds to hawk flights at various inland and coastal concentration points. But it was Broun (1951, 1963) who developed the additional concept that good hawk flights at Hawk Mountain, Pennsylvania, usually depend upon the passage of a low-pressure area across upper New York State and lower New England, followed by a cold front moving across the eastern United States or at least the region of the hawk flyways. In autumn, westerly or north-westerly winds generally occur after the passage of these fronts. Two or three days later, sometimes sooner, notable hawk flights usually appear along the mountains (particularly the Kittatinny Ridge) of Pennsylvania and New Jersey as well as at various coastal points. A similar relationship of general weather patterns to autumn hawk flights occurs at various Great Lakes concentration points (Haugh, 1972; Hofslund, 1966; Mueller and Berger, 1961).

Nonetheless, exceptions occur as Broun (1949a) points out in

Hawks Aloft. On 16 September 1948 at Hawk Mountain, Pennsylvania, for example, at least 11,392 hawks (mainly Broad-wings) were counted. Easterly winds prevailed. A similar flight of 10,101 hawks, again mainly Broad-wings, occurred on 19 September 1970 at the Montclair Hawk Lookout Sanctuary, New Jersey, as light winds prevailed from the northwest to the north to the northeast. At Cape Charles, Virginia, easterly winds also are more favorable for large hawk flights (Rusling, 1936). However, geography plays a critical role at this spot as hawks travel the length of a long, narrow peninsula before becoming concentrated at its southern tip.

The fact that not all major hawk flights conform to the classic weather-pattern correlation described by Broun (1951, 1963) suggests that other factors also influence hawk flights. Robbins (1956: 211), for example, concluded:

"In the New England States the heaviest flights of broad-winged hawks usually occur during the first one to three days after the passage of a cool front, while the wind is still from the northwest. During a period of warm, southwesterly air flow, on the other hand, few migrating broad-wings have been observed in the northeast.

"We used to think that the same conditions prevailed in the Middle Atlantic States. And we still expect a better flight on a northwest wind than on a southwest wind. But some of the greatest movements through Pennsylvania, Maryland and Virginia have been recorded when the wind was from the northeast, or when it was virtually calm. Flights have also been so high above ground that the birds were barely visible even with the aid of binoculars. The peak flight of broad-wings at Mount Tom in Massachusetts can frequently be predicted within one or two days by studying the weather maps. But by the time the flights have progressed as far south as the Middle Atlantic States, they no longer show such a close correlation with weather conditions, and it is difficult to predict when or where the most spectacular movements will take place.

"Apparently weather conditions are more important in initiating southward migration than they are in halting it once it has begun. These hawks will continue to fly no matter what quarter the wind is from, and they will even fly around scattered showers.

Wind direction and speed can be of the utmost importance to the observer, however. Light or moderate northwesterly winds on a cool day make favorable observation conditions on most of the Appalachian ridges. When winds are light and variable and the day is warm, there is a greater tendency for broad-wings to leave the ridges and move cross-country, by getting their 'lift' from thermals. There is some evidence that strong northwesterly winds cause migrating hawks to drift toward the coast."

To study the relationship of autumn hawk migrations to general weather systems—particularly in respect to the role of weather as a releaser of hawk flights—one must compare daily hawk counts with weather patterns shown on daily weather maps. This technique was refined by Broun (1951, 1963) in an analysis of the 1950 hawk flights at Hawk Mountain, Pennsylvania. Following is a similar analysis of one season at Bake Oven Knob, Pennsylvania, and another analysis of the seasonal hawk flights at several northeastern lookouts.

THE 1967 SEASON

Moderately good hawk flights occurred during August and early September, but it was not until 10 September that a notable flight developed. On 9 September a large high-pressure area covered the East coast as a cold front approached Pennsylvania from the Great Lakes region. On the 10th the front had crossed Pennsylvania. At Bake Oven Knob cool air temperatures and 20–30 mile-per-hour north-northwest winds prevailed, and hawks of eight species were seen, including 384 Broad-wings, 5 Bald Eagles, 37 Ospreys, and 15 Sparrow Hawks.

During the next seven days, 11 through 17 September—the traditional period of peak Broad-wing flights in eastern Pennsylvania—a huge high covered the East coast as hurricane Doria appeared in the Atlantic Ocean, approached the coast on the 14th and came ashore on the 16th in the area from Virginia to New Jersey.

On 11 September a poor flight occurred at Bake Oven Knob, but on the 12th a sizeable movement was recorded. Light east winds prevailed throughout most of the day with slightly cool air tem-

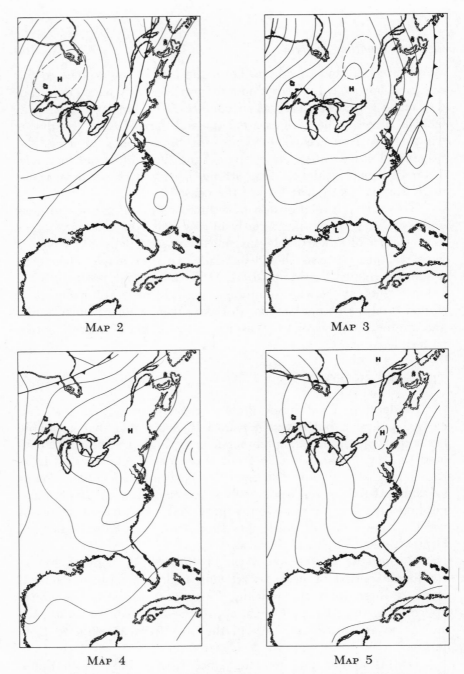

MAP 2   MAP 3

MAP 4   MAP 5

MAPS 2–5   Weather systems for the period 10–13 September 1967 correlated well with large flights of Broad-winged Hawks at Bake Oven Knob, Pa., which occurred during the next few days.

peratures. Seven species of hawks were counted including 1,844 Broad-wings, 1 Peregrine Falcon, and 10 Sparrow Hawks. The Broad-wing flight was seemingly without pattern. "Groups of birds appeared both north and south of the ridge, as well as overhead," I wrote in my notes. "It appears that a general broad movement of Broad-wings was drifting across country and using the Blue Mountain (Kittatinny Ridge) as a flight line at some locations and not at others. Radio communication with two Hawk Mountain lookouts confirmed this." Between 1400 and 1700 hours more than 1,000 Broad-wings passed. Many flew very low, and hundreds landed in the woods around our lookout.

On 13 September the season's peak Broad-wing flight developed at Bake Oven Knob despite the early date and the high-pressure area blanketing the East coast. During the morning north winds at 0 to 5 miles per hour prevailed, but they shifted to the southeast during the afternoon. Slightly cool air temperatures of the morning increased to comfortable levels in the afternoon. The flight in the morning was rather poor, but an hour after the wind shift to the southeast a massive Broad-wing flight developed with hawks appearing over valleys both north and south of the ridge. Again the flight appeared to be drifting cross-country. Despite our high count Hawk Mountain recorded relatively few birds. Radio communication with observers there showed that many Broad-wings were drifting toward the Pinnacle. The birds were using strong thermal activity. The hawk count at Bake Oven Knob included 14 Turkey Vultures, 1 Sharp-shin, 5 Red-tails, 2,575 Broad-wings, 2 adult Bald Eagles, 3 Marsh Hawks, 6 Ospreys, 1 Sparrow Hawk, and 2 unidentified hawks.

As the high continued to cover the East, small hawk flights occurred at the Knob from 14 through 18 September. Observations were not made on the 19th, but the next day east to southeast winds occurred at the Knob and 11 species were seen, including an adult Golden Eagle, 46 Ospreys, a Peregrine Falcon, a Pigeon Hawk, and 7 Sparrow Hawks.

On the 21st a cold front was approaching Pennsylvania and had crossed the state by the next day. Rain developed in central Pennsylvania on the 22d and in much of New York and New England. At Bake Oven Knob, however, northwest winds were recorded, but only 141 hawks of six species were counted. Poor hawking also developed on the 23d. Rain developed by the 25th, and counts were not made on the 25th or 26th.

On 27 September Pennsylvania was covered by a high-pressure area as a wet cold front approached from the Great Lakes region. Robert and Anne MacClay reported light to brisk southwest winds at Bake Oven Knob, and a fair hawk flight developed, including 207 Broad-wings, an adult Golden Eagle, and 24 Ospreys.

On 30 September observations were continued as light to brisk westerly winds developed, bringing 188 hawks to the Knob. Included were 114 Sharp-shins and 19 Sparrow Hawks. A cold front crossed Pennsylvania on the 30th, and scattered storms covered sections of New York and New England. Nonetheless, the flight continued through 1 October as the front was well off the coast, New York and New England were covered by a low and scattered rain, and Pennsylvania was divided by a low and a high. On the 1st at Bake Oven Knob, light to fairly brisk west winds occurred, and the following hawks were seen: 1 Turkey Vulture, 135 Sharp-shins, 3 Cooper's Hawks, 1 Red-tail, 35 Broad-wings, 2 Marsh Hawks, 18 Ospreys, 20 Sparrow Hawks, and 6 unidentified hawks.

Observations were then discontinued until 4 October when light west winds occurred at the Knob. A flight of 326 hawks passed including 246 Sharp-shins, 21 Ospreys, and 19 Sparrow Hawks. The flight seemed to be triggered by a low covering the East since 1 October and the passage of a cold front-occluded front across northern New York and New England.

Between 5 and 10 October only two days of observations were possible at Bake Oven Knob, and no notable hawk flights were seen. On the 11th, however, 138 Sharp-shins passed as light to moderate west to northwest winds prevailed. A wet low coupled with a complicated frontal system in northern New England and southeastern Canada probably caused this flight. On the 12th a huge high covered the East, and light northwest winds were recorded at the Knob. Of the 107 hawks counted, 79 were Sharp-shins—perhaps a carry-over flight from the previous day.

Another notable hawk flight did not develop until 16 October as a high covered the East coast. On the 16th, however, 236 hawks were seen at the Knob, including 178 Sharp-shins and 12 Marsh

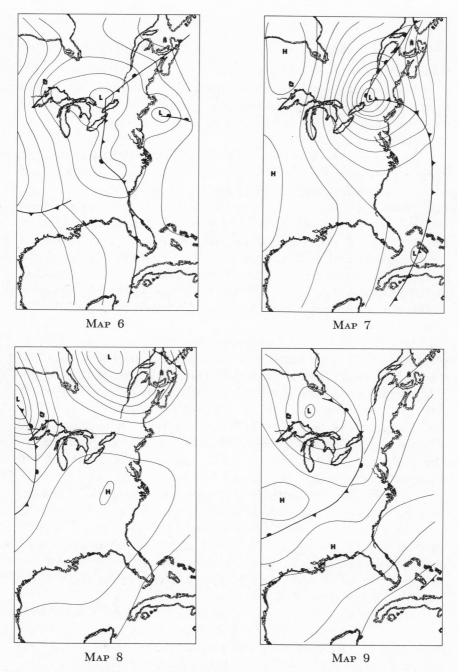

MAP 6   MAP 7

MAP 8   MAP 9

MAPS 6–9 Weather systems for the period 18–21 October 1967 again correlated well with good hawk flights at Bake Oven Knob, Pa. A low-pressure
area off New England on the 18th, followed by the passage of a cold front the
next day, triggered hawk migrations resulting in good hawk flights at the Knob
on 19 and 20 October. The passage of another cold front through the East
coast on 21 October produced another large hawk flight.

Hawks. Light to moderate southwest winds prevailed. On the 17th a wet stationary front extended from the eastern Great Lakes through northern New England as a low covered parts of Maine and New Brunswick. Few hawks were seen, and observations were not made on 18 October.

On 19 October, however, a massive cold front crossed Pennsylvania, New Jersey, southern New York, and southern New England while a warm front extended along the Saint Lawrence seaway. The entire northeast was centered in a low. At Bake Oven Knob moderate to brisk west winds developed, and 212 hawks, including 180 Sharp-shins, were counted. On the 20th a low covered northern New England and southeastern Canada as a high blanketed the remaining sections of the East. Weather data were not recorded at Bake Oven Knob, but at nearby Allentown light west to southwest winds were logged. Of the 216 hawks seen at the Knob, 174 were Sharp-shins.

At 0100 hours on 21 October an occluded front crossed Pennsylvania, and by 1300 hours it was classed as a cold front and had reached the coast. Observers at Bake Oven Knob noted light southwest winds during the morning and light to moderate west winds during the afternoon. Cool air temperatures also occurred. The day's hawk count included 5 Turkey Vultures, a Goshawk, 249 Sharp-shins, 9 Cooper's Hawks, 19 Red-tails, 2 Red-shoulders, 12 Marsh Hawks, 6 Ospreys, 2 Peregrine Falcons, 2 Pigeon Hawks, 16 Sparrow Hawks, and 2 unidentified hawks.

Poor flights were seen from 22 October through 1 November as a high covered the East although storms over New England, coupled with a cold front passing across the East on 25 and 26 October should have produced a good flight. A high over the East on the 27th gave way to a cold front-occluded front crossing the area on 28 October, and another high covered the region from the 29th of October through 1 November.

Limited hawk counting was done at Bake Oven Knob during early November. However, on 8 November, light to moderately brisk west to northwest winds developed at the Knob, and Robert and Anne MacClay counted 172 Red-tails, 6 Red-shoulders, 2 immature Golden Eagles, a Peregrine Falcon, and 1 unidentified hawk. The classical weather pattern involving a low and a cold front

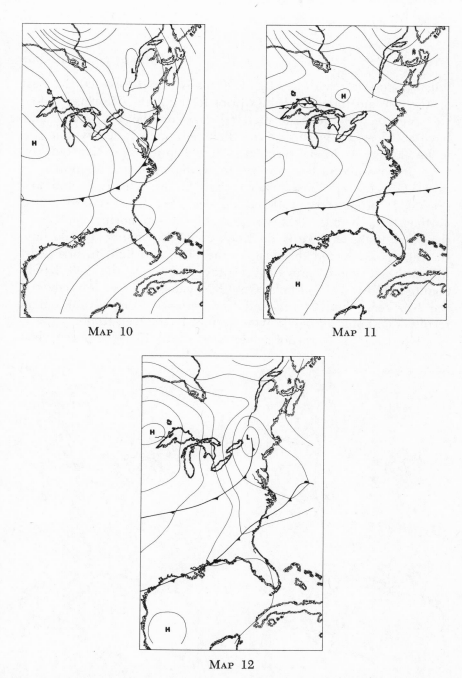

MAP 10

MAP 11

MAP 12

MAPS 10–12 Weather systems in the East for the period 13–15 November 1967 correlated well with hawk flights at Bake Oven Knob, Pa., on 15 November.

failed to correlate with this flight. Actually a high covered the East on the 8th and for several days prior to the flight. However, a low was centered off the New England coast on the 8th, and it may have acted as a releaser for the flight although it seems doubtful if the hawks could have reached eastern Pennsylvania that quickly.

On 9 November a low covered the Canadian maritime provinces as a high covered the eastern United States. The high and low prevailed through the 10th, as an occluded front passed across northeastern Canada. This was probably too far north to influence the hawk flight at Bake Oven Knob on 11 November. In addition, the high still covered the East on the 11th, and light to brisk east to southeast winds prevailed at the Knob. Of the 148 hawks counted, 136 were Red-tails. On 12 November a wet low covered the Northeast, and a wet cold front crossed Pennsylvania. Rain resulted, and only 12 hawks were seen at Bake Oven Knob.

Observations were not made again until 15 November, when

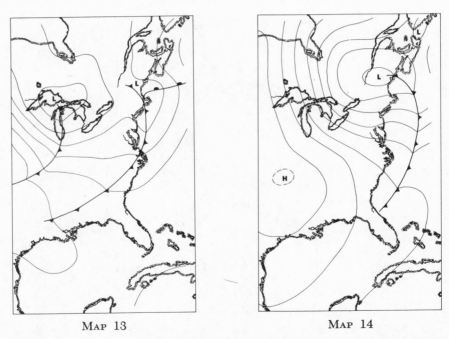

MAP 13   MAP 14

MAPS 13–14   On 18–19 November 1967, East coast weather systems again correlated well with good hawk flights at Bake Oven Knob, Pa., on 19 November.

a low covered New York and southern New England and a cold front crossed Pennsylvania. Very brisk west-northwest winds resulted. At Bake Oven Knob, Robert MacClay counted 4 Goshawks, 123 Red-tails, 5 Red-shoulders, 4 adult Golden Eagles, and 2 unidentified hawks.

Counts were not made again until 19 November when brisk west winds and very cold air temperatures occurred. A flight of 141 hawks of six species included 2 Goshawks, 2 Sharp-shins, 127 Red-tails, 2 Rough-legs, 1 adult Golden Eagle, 3 Marsh Hawks, and 4 unidentified hawks. The weather map for the 18th shows a large cold front having crossed the East with an enormous low behind the front. On the 19th the wet low still covered the Northeast as the cold front was off the coast. The season's hawk watch was terminated at Bake Oven Knob on 19 November.

THE 1968 SEASON

To examine the role of general weather systems to hawk migrations further, the 1968 season will be analyzed using data gathered at four northeastern hawk lookouts: Mount Peter, New York, Montclair Hawk Lookout Sanctuary, New Jersey, Bake Oven Knob, Pennsylvania, and Hawk Mountain Sanctuary, Pennsylvania. Data for the first two sites are taken from mimeographed reports issued by the coordinators of each station. Bake Oven Knob data are in my files, and the Hawk Mountain data are published in News Letter No. 41 of the Hawk Mountain Sanctuary Association for March 1969.

On 2 September a stationary front was off the Atlantic coast, and New England and the Middle Atlantic states were covered by a low-pressure area. From 3 through 6 September, however, huge highs covered the East as a cold front on the 6th passed across western New York and Pennsylvania. On the 7th the front was in the Atlantic just off the coast. At Mount Peter a flight of 43 hawks was recorded, mainly Broad-wings, and a similar count of 48 hawks was tabulated at Montclair, where southerly winds occurred. However, at Bake Oven Knob light west-northwest winds prevailed, and slightly cool air temperatures were recorded. Ten species numbering 58 hawks were seen, including a Goshawk (unusually early),

28 Broad-wings, and 1 adult Bald Eagle. At Hawk Mountain, on the other hand, 158 hawks of eight species were sighted, including 111 Broad-wings, 3 Bald Eagles, 1 Peregrine Falcon, and 30 Sparrow Hawks. Apparently the weather pattern had different effects upon birds of several different populations spread over a large geographic area.

On 8 September the East coast was covered by a large high with isolated rain in eastern Massachusetts and Rhode Island. Again watchers at Mount Peter counted only 74 hawks of which 57 were Broad-wings. Curiously, the same number of birds was seen at Montclair, but 22 were Broad-wings and 42 were Sparrow Hawks. Along Pennsylvania's hawk ridges good flights were seen, however. At Bake Oven Knob, light east winds developed, and 283 hawks of nine species were counted. Included were a Goshawk, 233 Broad-wings, 14 Marsh Hawks, and 12 Ospreys. At Hawk Mountain 606 hawks were seen, including 542 Broad-wings, a Bald Eagle, 13 Marsh Hawks, and 21 Ospreys.

Poor hawking occurred at all four lookouts from 9 through 11 September as a high, followed by rain, covered the Middle Atlantic states and a low developed in eastern Canada on the 11th and extended southward across New England until 14 September. However, on the 11th a cold front passed through the area and was off the coast on the 12th. All four lookouts experienced fairly good to good hawk flights on 12 September. At Mount Peter 671 birds were seen, including 575 Broad-wings and 86 Sparrow Hawks. At Montclair, where northwest winds prevailed, 133 hawks were seen of which 67 were Broad-wings and 49 were Sparrow Hawks. Meanwhile, in Pennsylvania, fairly brisk west-northwest winds and cool air temperatures were noted at Bake Oven Knob along with 356 hawks: 4 Turkey Vultures, 20 Sharp-shins, 4 Cooper's Hawks, 6 Red-tails, 250 Broad-wings, 9 Bald Eagles (6 adults and 3 immatures), 3 Marsh Hawks, 19 Ospreys, 39 Sparrow Hawks, and 2 unidentified hawks. At Hawk Mountain the flight was even larger. Observers there counted 825 hawks: 26 Sharp-shins, 2 Cooper's Hawks, 3 Red-tails, 689 Broad-wings, 6 Bald Eagles, 11 Marsh Hawks, 31 Ospreys, and 57 Sparrow Hawks.

Apparently the general weather systems prior to the 12th caused some hawk populations to initiate migrations. That different

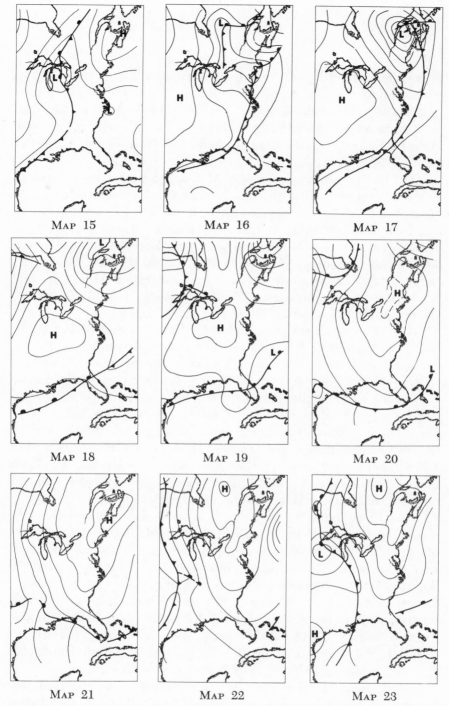

Maps 15–23 Active weather systems in the East on 10–12 September 1968 triggered northeastern hawk migrations which continued in varying degrees of magnitude at different New York, New Jersey, and Pennsylvania lookouts through 18 September.

populations of birds were involved is apparent from the analyses which follow for 13 through 19 September. It is important to note that during the whole of this seven-day period much of the East was continually blanketed by a huge high, although a low with associated rain covered eastern Canada and New England from 11 through 14 September.

On 13 September observers at Mount Peter counted 234 hawks, of which 189 were Broad-wings and 29 were Sparrow Hawks. At Montclair 1,330 hawks were counted, including 1,211 Broad-wings and 64 Sparrow Hawks. At Bake Oven Knob, where cool air temperatures and 0–15 mile-per-hour west winds prevailed, we counted 695 hawks, including 576 Broad-wings, 2 adult Bald Eagles, 14 Ospreys, and 38 Sparrow Hawks. And, at Hawk Mountain, 977 hawks were seen, including 778 Broad-wings, 2 Bald Eagles, 15 Ospreys, and 123 Sparrow Hawks.

On 14 September a poor flight (72 birds) passed Mount Peter, but a peak flight of 5,513 hawks was noted at Montclair. Broad-winged Hawks accounted for 5,371 birds and Sparrow Hawks for another 67 individuals. West to northwest winds prevailed. In eastern Pennsylvania, light to moderate west winds and cool air temperatures occurred at Bake Oven Knob, but only 698 hawks were seen. These included 610 Broad-wings, 1 adult Bald Eagle, and 18 Ospreys plus various other species. However, at Hawk Mountain 1,396 hawks were seen, with Broad-wings accounting for 1,265 birds. Additionally 3 Bald Eagles were sighted, along with 23 Ospreys, 43 Sparrow Hawks, and other species.

On 15 September hawk flights were equally interesting. Observers at Mount Peter recorded their peak flight—2,962 hawks of which 2,928 were Broad-wings. In contrast, Montclair reported only 204 hawks, including 98 Broad-wings, 22 Ospreys, and 60 Sparrow Hawks. At Bake Oven Knob calm to light east winds were logged along with 1,780 hawks. This total included 1,668 Broad-wings and 28 Ospreys. Curiously, watchers at Hawk Mountain saw 3,320 hawks—1 Goshawk, 36 Sharp-shins, 7 Cooper's Hawks, 2 Red-tails, 3,211 Broad-wings, 3 Bald Eagles, 19 Marsh Hawks, 29 Ospreys, 9 Sparrow Hawks, and 3 unidentified hawks.

On 16 September another memorable day occurred for observers stationed at three of the four lookouts. At Mount Peter,

1,628 hawks were seen, including 1,608 Broad-wings. But at Mont-clair only 18 birds (2 Broad-wings) were seen. On the Pennsylvania hawk ridges major flights were logged, however. At Bake Oven Knob calm to light and variable winds produced 2,534 hawks— 2,489 Broad-wings, 15 Ospreys, and various other species; 1,892 of the Broad-wings were seen between 1500 and 1600 hours. At Hawk Mountain observers counted 2,193 hawks, of which 2,094 were Broad-wings and 30 were Ospreys.

On 17 September the Broad-wing flights again were spec-tacular at three of the four lookouts. Mount Peter logged another large flight, this time numbering 1,049 birds, of which 1,030 were Broad-wings. The Montclair flight, on the other hand, was small. Only 50 birds were seen, and 6 of these were Broad-wings. In con-trast, hawk flights on the Pennsylvania ridges on the 17th were excellent. At Bake Oven Knob calm to light east to southeast winds prevailed, coupled with fairly warm air temperatures. Of the 4,220 hawks counted, 4,137 were Broad-wings. The majority of these passed during midafternoon. For example, 1,340 were seen between 1400 and 1500 hours, and 2,416 appeared between 1500 and 1600 hours. We also counted an adult Bald Eagle, 31 Ospreys, and a few other hawks. At Hawk Mountain, however, an even more impres-sive flight developed. A total of 5,359 hawks was tabulated, includ-ing 5,259 Broad-wings, a Bald Eagle, 34 Ospreys, and a handful of other birds.

On 18 September poor flights passed two of the lookouts, but moderate to excellent flights passed the other two sites. At Mount Peter only 85 hawks were seen, and at Montclair 17 hawks passed, including a single Broad-wing. But westward, along the Pennsyl-vania hawk ridges, substantially more birds were seen. At Bake Oven Knob comfortable air temperatures and light east winds pre-vailed during most of the day, and 972 hawks were seen. Broad-wings accounted for 900 of these. At Hawk Mountain a very large Broad-wing flight developed, however. Of the 3,041 hawks counted, 2,981 were Broad-wings. The geographic distribution of the Broad-wing flight on the 18th suggests that the bulk of the day's birds were moving cross-country and funneling out of the Poconos onto ridges immediately adjacent to and north of the Kittatinny Ridge. The birds then became concentrated as they reached points in the vicinity of Hawk Mountain, where the northern ridges converge and bend toward Hawk Mountain. Apparently the birds were then forced toward the Kittatinny, where they appeared in the general vicinity of Hawk Mountain in large numbers.

As the high continued to blanket the Middle Atlantic states from 19 through 27 September, poor hawking developed at each of the four lookouts. Fewer than 100 hawks were counted daily at Mount Peter and at Montclair, and counts in the low hundreds were made at Bake Oven Knob and at Hawk Mountain. However, an additional factor related to these poor counts is the fact that the peak Broad-wing season was past; the bulk of these birds already had passed through the region.

On 26 September a low covered northern New England as a cold front reached the Atlantic coast. On the 27th and 28th the front was off the coast as a high covered the Middle Atlantic states. At Mount Peter 26 hawks were seen, and at Montclair 72 birds of five species passed, including 30 Sharp-shins. In Pennsylvania, at Bake Oven Knob, observations were not made, but at Hawk Mountain 235 hawks of nine species passed, including 181 Sharp-shins.

Most of the United States was covered by a very large high on 29 September. At Mount Peter and at Montclair few hawks were seen, but light west to northwest winds and cool air temperatures at Bake Oven Knob resulted in a moderate flight of 149 hawks, 106 of which were Sharp-shins. At Hawk Mountain a larger flight of 222 hawks was noted, including 148 Sharp-shins.

Poor hawking continued on the 30th at Mount Peter and Montclair, and observations were terminated for those sites for the season. Bake Oven Knob was not covered, but at Hawk Mountain 138 hawks were counted, including 79 Sharp-shins.

On 1 and 2 October, highs covered the eastern United States as an occluded front was positioned west and north of the Great Lakes. At Hawk Mountain on the 1st, 199 hawks were seen, including 175 Sharp-shins. On 2 October south to southwest winds occurred at Bake Oven Knob along with warm air temperatures. The count totaled 246 birds, of which 179 were Sharp-shins. At Hawk Mountain 169 hawks were seen, of which 90 were Sharp-shins.

By 3 October the occluded front had changed to a cold front and was approaching western Pennsylvania bringing scattered areas of rain. Only 42 hawks were seen at Hawk Mountain. On the 4th the front was off the Atlantic coast, and most of the country was covered by an enormous high. The count at Hawk Mountain was 245 birds, 190 of which were Sharp-shins.

The high continued to blanket the East on the 5th and 6th, and good hawk flights developed. Cool air and moderately brisk northwest winds occurred at Bake Oven Knob on the 5th, bringing 207 Sharp-shinned Hawks and various other species for a total count of 279. At Hawk Mountain 483 hawks passed, of which 371 were Sharp-shins. Equally good hawk flights continued through the 6th despite light to moderate southeast winds and cool air. At the Knob, observers counted 374 hawks of ten species, including 257 Sharp-shins, 19 Marsh Hawks, 26 Ospreys, and 4 Peregrine Falcons. On 7 October observers at Hawk Mountain saw 373 hawks despite rain over parts of Pennsylvania and New England.

Poor flights occurred at Bake Oven Knob and Hawk Mountain from 8 through 12 October as various low- and high-pressure areas, and one wet cold front (on the 10th and 11th), crossed the Middle Atlantic states. On 13 October, however, observers at Bake Oven Knob and Hawk Mountain noted moderately good flights, followed by rain on 14 October as a cold front pushed down into Pennsylvania from New England. On 15 October, most of the East was covered by a large high although a belt of precipitation extended across upper Maine and most of New Brunswick. Hawk watchers at Bake Oven Knob noted light southwest winds and counted 2 Goshawks, 220 Sharp-shins, 1 Cooper's Hawk, 11 Red-tails, 16 Marsh Hawks, 1 Osprey, and 10 unidentified hawks. Southwestward along the Kittatinny Ridge, at Hawk Mountain, 162 hawks were seen, including 1 Goshawk, 142 Sharp-shins, 2 Cooper's Hawks, 3 Red-tails, 1 Red-shoulder, 1 Broad-wing, 9 Marsh Hawks, 2 Ospreys, and 1 Pigeon Hawk.

The high continued to cover the East through 17 October. On the 16th at Bake Oven Knob, very light westerly and southwesterly winds prevailed as 236 hawks of nine species were counted, including 117 Sharp-shins, 72 Red-tails, 12 Marsh Hawks, and 11 Sparrow Hawks. At Hawk Mountain 254 hawks, including 147 Sharp-shins, 38 Red-tails, 12 Marsh Hawks, and 8 Sparrow Hawks, were counted. Good hawking continued through 17 October, when light southwest winds prevailed at Bake Oven Knob and nine species of hawks were seen, including 4 Goshawks, 117 Sharp-shins, 31 Red-tails, 1 Golden Eagle, and 13 Marsh Hawks. Ten species were seen at Hawk Mountain on the 17th, including 1 Goshawk, 99 Sharp-shins, 37 Red-tails, 1 Golden Eagle, 18 Marsh Hawks, 1 Peregrine Falcon, and 3 Pigeon Hawks.

Rain covered eastern Pennsylvania on 18 and 19 October as a cold front crossed eastern Pennsylvania. By 0700 hours on 20 October the cold front was off the coast as hurricane Gladys was positioned along coastal North Carolina and Virginia. Light to moderate northwest to west winds coupled with cool air temperatures were noted at Bake Oven Knob, but only 123 hawks were counted. At Hawk Mountain 239 hawks were seen.

On 21 October excellent hawk flights were seen in eastern Pennsylvania. At Bake Oven Knob brisk northwest winds and cool air temperatures brought 7 Turkey Vultures, 6 Goshawks, 55 Sharp-shins, 2 Cooper's Hawks, 258 Red-tails, 5 Red-shoulders, 1 Bald Eagle, 19 Marsh Hawks, 1 Osprey, 1 Pigeon Hawk, 3 Sparrow Hawks, and 6 unidentified hawks. At Hawk Mountain 7 Goshawks, 104 Sharp-shins, 6 Cooper's Hawks, 253 Red-tails, 21 Red-shoulders,

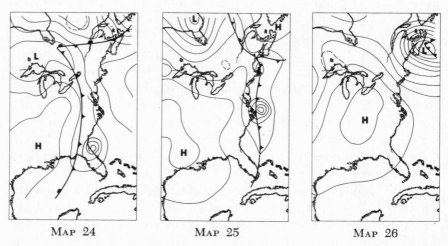

MAP 24   MAP 25   MAP 26

MAPS 24–26   The passage of a wet cold front across the East coast on 19 October 1968 produced good to excellent hawk flights at Bake Oven Knob and Hawk Mountain, Pa., on 20–21 October.

2 Golden Eagles, 1 Bald Eagle, 18 Marsh Hawks, 2 Pigeon Hawks, 6 Sparrow Hawks, and 2 unidentified hawks were counted.

By 22 October a cold front was crossing the western Great Lakes, a low-pressure area was centered above New Brunswick, and a high still covered the Middle Atlantic states. Moderately good hawk flights passed both Bake Oven Knob and Hawk Mountain. On the 23d the cold front crossed Pennsylvania, and a flight of 139 hawks was seen at Hawk Mountain. Bake Oven Knob was not covered that day. On the 24th the front was stationary off the coast, and poor hawking occurred in eastern Pennsylvania. By 25 October rain covered the eastern Great Lakes as a low was centered off upper New England, and the stationary front still remained off the coast. Observations were not made at Bake Oven Knob, but at Hawk Mountain 160 hawks were seen.

On 26 October a low extended across upper New England as a high covered the rest of the East coast. Cool air temperatures and moderate to brisk northwest to west winds prevailed at Bake Oven Knob, where 224 hawks were seen: 5 Turkey Vultures, 5 Goshawks, 35 Sharp-shins, 161 Red-tails, 4 Red-shoulders, 2 Golden Eagles, 10 Marsh Hawks, 1 Sparrow Hawk, and 1 unidentified hawk. At Hawk Mountain 213 hawks were counted, including 7 Goshawks, 27 Sharp-shins, 2 Cooper's Hawks, 162 Red-tails, 6 Red-shoulders, 1 Broad-wing, 1 Bald Eagle, 5 Marsh Hawks, 1 Peregrine Falcon, and 1 Sparrow Hawk.

On 27 October the low still covered upper New England, and the high continued to blanket the rest of the East coast as a cold front crossed the western Great Lakes. Cool air temperatures again were recorded at Bake Oven Knob as moderately light westerly winds shifted to the southwest. We counted 202 hawks of eight species, including 118 Red-tails and 30 Red-shoulders. At Hawk Mountain 228 hawks passed, including 142 Red-tails and 26 Red-shoulders.

A cold front crossed Pennsylvania on 28 October as a large, wet low was centered over the Great Lakes. Calm to very light and variable winds prevailed at the Knob, resulting in a count of 147 hawks. Only 74 hawks passed Hawk Mountain. On the 29th the cold front was far off the Atlantic coast as the low still blanketed the East. Cool air temperatures, coupled with moderate to brisk west winds, were noted at Bake Oven Knob. We counted 144 hawks whereas Hawk Mountain logged 173 hawks.

The full influence of the weather systems of the preceding days was not felt on the Pennsylvania hawk ridges until 30 October, however. Then the low still covered New England and a high extended into Pennsylvania. Moderate northwest to west winds and cool air temperatures occurred at Bake Oven Knob, and the count of 454 hawks included 10 Goshawks, 8 Sharp-shins, 3 Cooper's Hawks, 350 Red-tails, 69 Red-shoulders, 8 Marsh Hawks, and 5 unidentified hawks. The count of 503 hawks at Hawk Mountain included 11 Goshawks, 17 Sharp-shins, 6 Cooper's Hawks, 436 Red-tails, 19 Red-shoulders, 7 Marsh Hawks, 1 Sparrow Hawk, and 6 unidentified hawks. Observations were not made at the Knob on the 31st, but the high covered the Middle Atlantic states and the low covered much of New England. Another excellent flight passed Hawk Mountain. The count of 481 included 23 Goshawks, 378 Red-tails, 44 Red-shoulders, and 2 Golden Eagles.

On 1 November another excellent flight occurred at Hawk Mountain, where 415 hawks were counted. Included were 3 Goshawks, 19 Sharp-shins, 3 Cooper's Hawks, 312 Red-tails, 51 Red-shoulders, 1 Golden Eagle, 20 Marsh Hawks, 1 Peregrine Falcon, 1 Pigeon Hawk, 1 Sparrow Hawk, and 4 unidentified hawks. Incomplete data are available from Bake Oven Knob for that date, however. The flight continued, although somewhat diminished in magnitude, on 2 November as light to moderate west winds and cool air temperatures prevailed at Bake Oven Knob, where 2 Turkey Vultures, 1 Goshawk, 13 Sharp-shins, 129 Red-tails, 1 Red-shoulder, 2 Golden Eagles, 8 Marsh Hawks, and 1 unidentified bird were counted. Downridge at Hawk Mountain, observers counted 6 Goshawks, 27 Sharp-shins, 3 Cooper's Hawks, 195 Red-tails, 6 Red-shoulders, 2 Golden Eagles, 15 Marsh Hawks, 1 Pigeon Hawk, and 2 unidentified hawks. Rain developed the next day as a cold front crossed eastern Pennsylvania, and relatively poor flights occurred at the Knob and at Hawk Mountain on 4 November. Rain and/or poor hawk flights occurred again in eastern Pennsylvania during the period 5 to 8 November. A wet low-pressure area covered New England on 8 November as a cold front crossed the East coast and was offshore. This weather system apparently triggered another hawk flight. On 9 November, 212 hawks passed Bake Oven Knob as cold air temperatures and light to moderate west winds prevailed. Included were 1 Turkey Vulture, 8 Goshawks, 10 Sharp-shins, 127 Red-tails, 54 Red-shoulders, 1 Rough-leg, 6 Marsh Hawks, and 5 unidentified hawks. At Hawk Mountain 174 birds were seen—8 Goshawks, 5 Sharp-shins, 2 Cooper's Hawks, 118 Red-tails, 21 Red-shoulders, 2 Marsh Hawks, and 18 unidentified hawks.

Observations were not made at Bake Oven Knob on 10 November, and only 25 hawks were seen at Hawk Mountain that day. Poor flights occurred at both stations on the 11th, and a snow storm covered the area on 12 November. Incomplete observations were made after the 12th at Bake Oven Knob, but observations continued daily at Hawk Mountain through the end of the month. On 13 November, for example, as the low from the storm still was centered over New England, observers at Hawk Mountain counted 289 hawks —mainly Red-tails. The next day 427 hawks were seen, including 24 Goshawks, 12 Sharp-shins, 375 Red-tails, 9 Red-shoulders, 1 Rough-leg, 1 Golden Eagle, and 5 Marsh Hawks. Moderate to fairly large flights occurred again at Hawk Mountain on 19, 20, 22, 23, and 25 November. Most of these flights again were more or less associated with the typical weather pattern of low-pressure areas in New England and cold fronts passing the vicinity of the hawk lookouts.

12

Hawk Migrations and Local Weather Variables

Haugh's (1972) analysis of data from Hawk Mountain, Pennsylvania, suggested that relatively weak statistical correlations exist between hawk counts and local weather variables. Nonetheless, field studies conducted at Hawk Mountain (Broun, 1935, 1939, 1949a) and Bake Oven Knob, Pennsylvania (Heintzelman and Armentano, 1964; Heintzelman and MacClay, 1972, 1973), coupled with similar studies conducted at a large number of other eastern hawk lookouts, clearly suggest that certain local weather variables can influence daily hawk flights. The following seem to be especially important: (1) wind direction, (2) wind velocity, (3) air temperature, (4) cloud cover, and (5) visibility.

WIND DIRECTION

Wind direction long has been known as an important variable affecting daily hawk flights. For example, many early workers recognized the influence of northwesterly winds upon hawk flights both at coastal and inland concentration points (Allen and Peterson, 1936; Broun, 1935, 1939, 1949a; Ferguson and Ferguson, 1922; Poole, 1934; and Trowbridge, 1895). More recent studies of Kittatinny

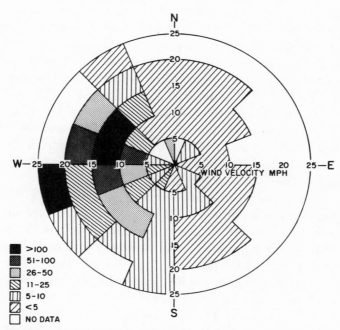

FIGURE 8   Correlation of Sharp-shinned Hawk migration with wind direction and velocity at Cedar Grove, Wis. The wind data are the vector means of the twelve hourly readings taken between 0600 and 1700 hours at Milwaukee, Wis. Radial lines enclose 22.5° increments of mean wind vector-directions (e.g., all mean winds between west and west-northwest). Concentric circles represent 5-mile-per-hour increments of the vector-mean velocity. The intensity of shaded segments indicates the mean number of Sharp-shinned Hawks observed per day under the indicated wind direction and velocity conditions. Redrawn from Mueller and Berger (1967b).

Ridge hawk flights generally confirm the importance of northwesterly winds (Broun, 1949a; Heintzelman and Armentano, 1964; Heintzelman and MacClay, 1972, 1973; Tilly, 1972a, 1972b, 1973).

At Hawk Mountain, Pennsylvania, for example, much attention has been given to the influence of wind upon daily hawk flights. Edge (1940a, 1940b) commented on the necessity of having strong winds, especially from northerly quarters, to keep hawks concentrated along the ridge. Southerly winds, she pointed out, tend to scatter the birds (Edge, 1939). Broun (1935, 1939, 1949a), in more detailed studies, determined that winds from almost any direction

can bring hawks to the Sanctuary but a brisk northwest wind is ideal
for the occurrence of the best flights. Strong updrafts are created,
and the hawks fly relatively close to the north slope of the ridge.
However, the opening of an alternate South Lookout about one-
half mile south of the main fold of the ridge has demonstrated, according to Nagy (1967) and Brett and Nagy (1973), that large
numbers of hawks drift southward, away from the ridge, when winds
with easterly or southerly components occur. In itself this fact was
well known (Broun, 1949a, 1961), but the magnitude of the numbers of birds which drift from the ridge, and are undetected from
the North Lookout, was not realized.

In a more sophisticated attempt to correlate Hawk Mountain
hawk flights with wind direction, Haugh (1972) demonstrated that
Red-tail and Sparrow Hawk flights correlated well with northerly
and westerly winds as did all hawk species combined. Red-shoulders seemed to occur in significantly greater numbers when northerly winds prevailed, westerly winds appearing to be less important.
Marsh Hawks did not show significant correlations with any wind
directions. Broad-winged Hawks showed good correlations with
northerly winds although more birds appeared on northeasterly
rather than northwesterly winds. Broad-wing data are subject to
considerable bias, however, since a few large flights can create
atypical situations. For example, Haugh (1972: 29) points out that
between 1954 and 1968, at Hawk Mountain, fifteen flights of over
2,000 Broad-wings per day occurred on northeasterly winds, and
two of these involved over 5,000 hawks (unusual for this site). In
contrast, only three flights in excess of 2,000 hawks per day occurred
on northwesterly winds during this period. Haugh's (1972) analysis
of these data revealed no statistically significant difference in respect
to the influence of northwesterly and northeasterly winds on Broad-
wing migrations. Sharp-shinned Hawk flights correlated best on
westerly winds.

At Bake Oven Knob, Pennsylvania, wind direction is equally
important in affecting hawk flights. Here the largest Sharp-shin
flights tend to occur on southwesterly to northwesterly winds although some good counts occasionally occur on southeasterly winds.
The Knob's largest Red-tailed Hawk flights almost always occur on

northwesterly winds, however. On the other hand, the largest Broad-winged Hawk flights tend to occur either on northwesterly or on east to southeasterly winds.

To illustrate the correlation of wind direction with hawk counts at Bake Oven Knob, Pennsylvania, I have plotted (figs. 9–11) Sharp-shin, Red-tail, and Broad-wing data for the period 1962–1971. Only flights numbering 100 or more birds, observed during the peak migratory periods for the respective species, are used for Sharp-shinned and Red-tailed Hawks. Similar limitations apply to the Broad-wing data except that flights of 400 or more birds are plotted.

At Raccoon Ridge, New Jersey, about fifty-five air miles northeast of Hawk Mountain, and also elsewhere on the Kittatinny Ridge,

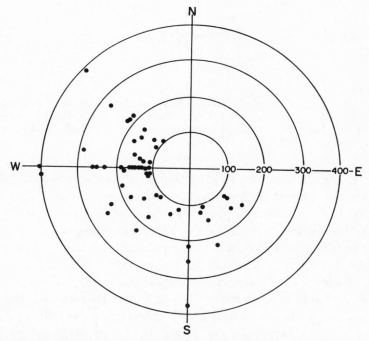

FIGURE 9   Correlation of Sharp-shinned Hawk flights (100 or more birds per day) and mean surface wind direction at Bake Oven Knob, Pa., for the period 1962 through 1971. The concentric circles represent increments in the numbers of birds counted.

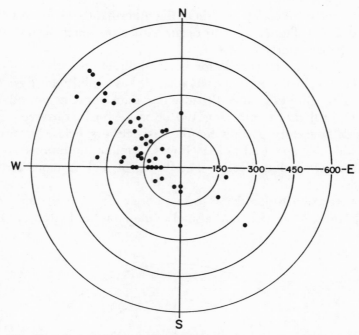

FIGURE 10   Correlation of Red-tailed Hawk flights (100 or more birds per day) and mean surface wind direction at Bake Oven Knob, Pa., for the period 1962 through 1971. The concentric circles represent increments in the numbers of birds counted.

Tilly (1972a, 1972b) determined that northwesterly winds tended to bring the largest flights although some good counts also were made on days with winds prevailing from other directions.

In the northeast, between the Kittatinny Ridge and the coast, there are a few lookouts on isolated mountains which generally do not extend for long distances. At Mount Peter, New York, for example, Bailey (1967) determined that northwest, west, or west-southwest winds at moderate to strong velocities generate the best hawk flights. Northwest winds tend to correlate especially well with many of Mount Peter's largest hawk flights. In northeastern New Jersey, the Montclair Hawk Lookout Sanctuary also is located on an isolated secondary mountain. There, too, local northwest winds invariably produce the best hawk flights (Andrew Bihun, Jr., verbal communication).

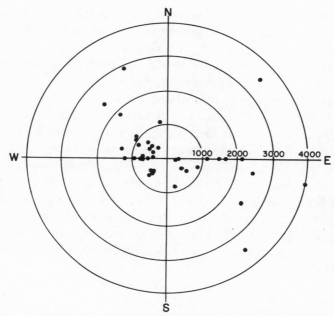

FIGURE 11 Correlation of Broad-winged Hawk flights (400 or more birds per day) and mean surface wind direction at Bake Oven Knob, Pa., for the period 1962 through 1971. The concentric circles represent increments in the numbers of birds counted.

Wind directions at coastal hawk lookouts also tend to favor the northwest for the largest flights. This applies to Fishers Island, New York (Ferguson and Ferguson, 1922), New Haven, Connecticut (Trowbridge, 1895), and Cape May Point, New Jersey (Allen and Peterson, 1936). But at Cape Charles, Virginia, the largest hawk flights occur when *northeasterly* winds prevail (Rusling, 1936) although local geography greatly influences hawk flights at this site.

WIND VELOCITY

Wind velocity also can exert a marked influence upon hawk flights. At Hawk Mountain, Pennsylvania, for example, during periods of light and variable winds, or even calms, Broad-winged Hawks leave the ridge and spread over broad migratory fronts (Broun, 1946: 3, 1960: 7; Edge, 1940a: 1, 1940b: 1). My field studies show that a similar effect occurs at nearby Bake Oven Knob. Indeed, at both locations it is fortuitous if large hawk flights occur as ridge flights when light winds, or calms, prevail.

Most hawks use the ridge flyway, or portions of it, when brisk winds create favorable updrafts. At Bake Oven Knob, for example, wind velocities ranging from about 5 to 15 or 20 miles per hour often produce ideal flight conditions. However, Broad-winged Hawks frequently find it difficult to remain aloft in high wind velocities. Indeed, I once watched a Broad-wing nearly buffeted into the rocks at Bake Oven Knob in unusually strong winds.

On the other hand, Red-tails, Golden Eagles, Bald Eagles, and Ospreys find brisk winds helpful. Often the best chances of seeing large flights of these birds occur during periods of high wind velocities. Ospreys, for example, occasionally reach speeds of 80 miles per hour under exceptionally favorable conditions (Broun and Goodwin, 1943).

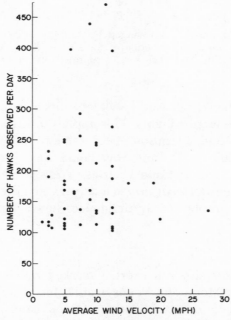

FIGURE 12   Correlation of Sharp-shinned Hawk flights (100 or more birds per day) and mean surface wind velocity at Bake Oven Knob, Pa., for the period 1962 through 1971.

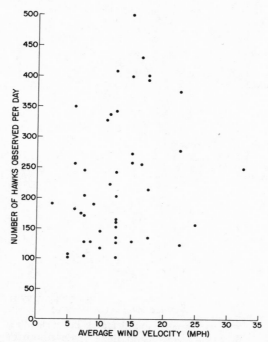

FIGURE 13   Correlation of Red-tailed Hawk flights (100 or more birds per day) and mean surface wind velocity at Bake Oven Knob, Pa., for the period 1962 through 1971.

The scatter diagrams (figs. 12–14) illustrate the relationship between Bake Oven Knob hawk counts and average daily wind velocity. The data cover the period 1962–1971. Only flights numbering 100 or more birds, recorded during the peak migration period for the species discussed, are used for the Sharp-shinned and Red-tailed Hawks. Similar conditions apply to the Broad-wing data with the exception that only flights of 400 or more birds are plotted.

AIR TEMPERATURES

The significance of correlations between air temperatures and hawk counts remains poorly understood. In fact, it may be of minor importance in most instances. Haugh (1972: 30–31), for example, concluded that ". . . a clear association between temperature and numbers of migrating hawks is not readily apparent for most spe-

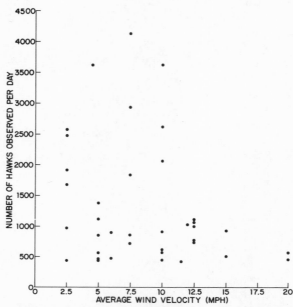

FIGURE 14   Correlation of Broad-winged Hawk flights (400 or more birds per day) and mean surface wind velocity at Bake Oven Knob, Pa., for the period 1962 through 1971.

cies." Since, for most species, the best correlations are associated with cool or cold air temperatures, which in turn are associated with the passage of cold fronts, the correlation of hawk counts with air temperature may merely reflect an association with frontal movements rather than with air temperature per se. Or such correlations may reflect the influence which the brisk winds associated with frontal movements, and the resulting updrafts, have upon migrating hawks.

The Broad-winged Hawk probably is an important exception. Along the Kittatinny Ridge, Broad-wing flights usually occur under two different types of weather conditions. Good flights frequently occur soon after the passage of cold fronts which, in turn, are associated with cool air temperatures and brisk northwesterly winds. Considerable use is made of the favorable updrafts along the ridge.

On the other hand, large Broad-wing flights also are seen several days after the passage of cold fronts through the area. By then

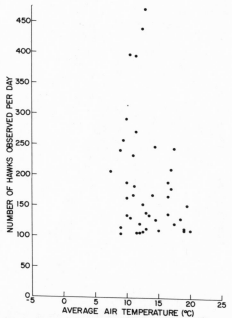

FIGURE 15 Correlation of Sharp-shinned Hawk flights (100 or more birds per day) with mean surface air temperature at Bake Oven Knob, Pa., for the period 1962 through 1971.

the influence of the front usually has diminished greatly, winds have shifted toward the east or south, and a warming trend develops. In September, it frequently becomes very hot on the lookouts as air temperatures reach the eighties or nineties. Moreover, wind velocities often are greatly reduced, or calms develop. Strong thermal activity develops under blue skies. These are ideal conditions for large cross-country Broad-wing flights. The hawks merely drift across the landscape, riding and gliding from thermal to thermal which, in turn, are produced by the intense sunlight striking the land. Updrafts along the ridges are used for limited distances only. Hence the correlation between Broad-wing counts and air temperature may be meaningful. Nonetheless, it could be argued that these correlations are, in fact, a relationship between birds and thermal production and use rather than air temperature per se.

The scatter diagrams (figs. 15–17) illustrate the relationship between air temperature and hawk flights at Bake Oven Knob, Penn-

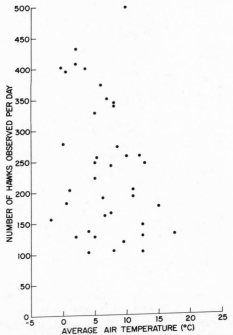

FIGURE 16 Correlation of Red-tailed Hawk flights (100 or more birds per day) with mean surface air temperature at Bake Oven Knob, Pa., for the period 1962 through 1971.

sylvania, for the period 1962–1971. Criteria similar to those used for selecting flights illustrated in the wind direction and wind velocity scatter diagrams were used here.

CLOUD COVER

Much remains to be learned about the migration of birds in or between opaque cloud layers, but Griffin (1973) provides an excellent summary of currently available information. Although Forbes and Forbes (1927) observed a flight of hawks, apparently Broadwings, enter and disappear into the opaqueness of clouds over Mount Monadnock, New Hampshire, and I observed similar events at Bake Oven Knob, Pennsylvania, I failed to understand the importance of cloud cover upon Broad-wing migrations until about

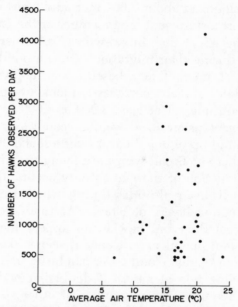

FIGURE 17 Correlation of Broad-winged Hawk flights (400 or more birds per day) with mean surface air temperature at Bake Oven Knob, Pa., for the period 1962 through 1971.

the midpoint in my field studies. Details of the significance of clouds upon Broad-wing migrations are discussed in chapter 15.

Briefly, during the September Broad-wing season, one sometimes notices on cloudless days that hawks rise increasingly higher during the day. By noon the birds frequently are nearly invisible or are pepper specks against the zenith even when observed through binoculars. These birds can easily pass undetected, creating an artifact of observation (and a serious bias in the day's count) referred to as the noon lull.*

Little accurate information is available regarding the maximum altitudes which Broad-wings, and other species, reach during migration. Stearns (1948: 5) observed Broad-wings at 3,000 feet over New Jersey from a blimp, Forbes and Forbes (1927) estimated

* An apparent halt in the migration for an hour or more around noon. No hawks are observed during this period. The phenomenon is discussed in detail in chapter 15.

hawks entering clouds at about 7,000 feet, and Murton and Wright (1968: 112) report a statement by a member of the Canadian Wildlife Service to the effect that Broad-wings "ride thermals up to at least 10,000 feet during their migrations south around the west end of Lake Ontario." I am confident, based upon my own observations of Broad-wing flights at many northeastern lookouts, that these birds routinely reach altitudes of at least 3,000 or 4,000 feet.

However, on occasion, local weather conditions cause a rapid partial or total build-up of cloud cover over a lookout. If this occurs when large numbers of Broad-wings are flying very high overhead, and a marked reduction of thermal activity occurs, the hawks may be forced to much lower altitudes during which time many birds can enter the opaque layers of clouds. At times, as described in chapter 15, several thousand hawks can appear flying out of the bottom of the cloud opaqueness. Hence it seems likely, at least for Broad-winged Hawks, that cloud cover can be an important variable influencing the detection of hawks. I doubt if cloud cover has an important influence upon the migrations of other species of hawks normally seen in the East, however. Haugh (1972) shares this viewpoint.

VISIBILITY

Under unusually severe conditions of local atmospheric haze and/or air pollution, these factors might seriously affect the accuracy of hawk counts by preventing many birds from being seen or by preventing correct identification. Detailed data are unavailable on this factor, however.

13

Hawk Migrations and Mountain Updrafts

Some of eastern North America's finest hawk concentration points are located on mountain ridges (others are located along coastal areas and along the Great Lakes). For example, thousands of hawks annually pass Hawk Mountain and Bake Oven Knob, Pennsylvania, and Raccoon Ridge, New Jersey—three notable points on the crest of the Kittatinny Ridge.

Why do large numbers of hawks follow ridges during their autumn migrations? Indeed, do hawks follow ridges such as the Kittatinny for extended distances or merely use them for varying, but generally limited, distances? Complete answers to these questions still require study. Nonetheless, the general mechanics of autumn hawk migrations along mountain ridges are known.

MOUNTAIN UPDRAFT FORMATION

In autumn, prevailing westerly surface winds strike the sides of ridges and are deflected upward, creating powerful updrafts. Haugh (1972: 39) also points out that lee waves * develop down-

---

* The undulatory movement of air downwind from an obstacle such as a mountain. Pennycuick (1973) reports that diurnal raptors use lee waves in East Africa although not much is known about this behavior.

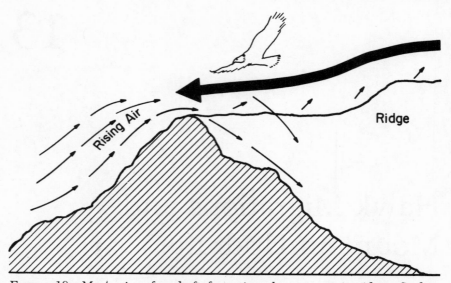

FIGURE 18   Mechanics of updraft formation along mountain ridges. Surface winds strike the sides of mountains and are deflected upward, thereby creating favorable air currents upon which migrating hawks soar. Redrawn from Heintzelman (1972d).

wind from mountain ridges but that the importance of these air currents to migrating hawks is unknown.

Most hawks and eagles are soaring birds. They are well adapted to using updrafts. Hence ridges and mountain chains—particularly those such as the Kittatinny, which extends virtually unbroken for many miles—provide natural flyways along which soaring hawks and eagles can travel relatively effortlessly. Broun (1935, 1939, 1949a) early recognized the basic importance of updrafts along mountains as a significant influence upon migrating hawks. However, in discussing the flights at Hawk Mountain, it was implied that hawks using the Kittatinny Ridge flyway do so for extended distances under ideal conditions. In many instances this may not be correct (see chapter 16). Edge (1940a), for example, noted that "Concentrated ridge flights take place only when strong northerly winds blow against Hawk Mountain, otherwise the flights are broken up and scattered." She also stated that without favorable winds striking against the sides of the mountain, thereby creating strong updrafts, Broad-winged and Sharp-shinned Hawks do not migrate along the ridge. Rather, they pass southward in a broad, scattered flight (Edge, 1940b). Broun (1949a) also was well aware of this. Despite these exceptions, the updrafts which exist along ridges, and the excellent flight opportunities they provide for most diurnal birds of prey, are major reasons why places such as Hawk Mountain and Bake Oven Knob are important hawk migration lookouts.

SOARING FLIGHT

Most of the species comprising the autumn hawk flights at mountain lookouts are accipiters, buteos, eagles, harriers, and Ospreys. These are all species which are more or less dependent upon soaring flight. Falcons, which are less dependent upon soaring on mountain updrafts, are much more abundant as coastal migrants (Allen and Peterson, 1936). It seems worthwhile, therefore, to examine the nature of soaring flight which can be divided into two types: (1) dynamic and (2) static. Static soaring further can be divided into several types (Cone, 1961, 1962b).

DYNAMIC SOARING

In theory, a variety of birds soaring over land could use dynamic soaring.* In practice, only high aspect-ratio birds such as albatrosses and similar seabirds are dependent upon this type of soaring (Cone, 1961). Cone points out that there is no average vertical velocity of air currents relative to the earth in dynamic soaring. Flight is dependent upon the nonuniformities of the wind in relation to time and location in space. Cone (1962b) further points out that the wind at sea level, and just above it, is slowed by friction which results in an increase in wind speed through the first 50 to 100 feet above the surface of the water. Albatrosses obtain the energy for their flight from this gradient in wind velocity. The birds glide downwind and, in so doing, convert potential energy into kinetic energy, thereby increasing both their airspeed and ground-speed.

When they are just above the waves they turn sharply into the

---

* Raspet (1960) defines dynamic soaring as "flight in an air mass which is not uniform in velocity."

wind, securing initial lifting acceleration as a result of an abrupt increase in their angle of attack. As the birds rise, they encounter increasingly higher wind speeds. Their airspeeds are thereby maintained which, in turn, allow them to rise again to the altitudes from which they originally descended in a glide without expending any energy of their own. Thus albatrosses and other seabirds can travel on motionless wings for hours utilizing dynamic soaring. A related type of dynamic soaring is rarely used by land birds by taking advantage of changing horizontal velocities in gusts of wind (Cone, 1962b).

STATIC SOARING

Migrating hawks frequently use some type of static soaring * which, according to Cone (1961), can be classified according to three ways in which vertical air currents are produced: (1) via declivity currents (referred to here as mountain updrafts), which result from air being diverted or deflected upward after striking a surface obstruction such as a mountain ridge, (2) via thermal currents caused by convection of buoyant fluid masses, or (3) via combinations of updrafts and thermals whereby unstable stratified air is set into buoyant convection by declivity motion initiated as the air passes over an obstacle such as a mountain ridge.

When migrating hawks engage in static soaring, using strong updrafts along mountain ridges, they attempt to adjust their shape to achieve as streamlined a condition as possible. This is achieved by folding back, i.e., not extending, their primary wing feathers and by keeping their tail feathers closed or pulled together rather than spreading them fan-shaped. Ospreys are particularly adept at streamlining their shape, resulting in their characteristic crooked-wing appearance when soaring on strong updrafts. The other diurnal raptors also are adept at streaming, but the effect is less dramatic than in the Osprey. Occasionally, during periods with extremely high wind velocities, a hawk will fold and compress its wings against its body almost completely. The bird thus virtually becomes a projectile hurtling through the air in what essentially is diving flight.

* Raspet (1960) defines static soaring as "flight in an air mass which has a vertical component."

Hence, by taking advantage of updrafts along mountain ridges, and using one of the forms of static soaring, many species of hawks can migrate for varying distances while using little or no energy.

RAPTOR USE OF MOUNTAIN UPDRAFTS

It is common knowledge that certain mountain ridges, notably the Kittatinny, which are oriented in a northeast to southwest direction, are major hawk flyways (Broun, 1949a; Heintzelman, 1972d). An examination of the species composition of the ridge flights can therefore be useful.

Using data provided by Broun (1949a: 151), one finds that four *Buteo* species represent about 68.11 percent of the hawks seen at Hawk Mountain, and three *Accipiter* species form another 26.29 percent. Combined, these seven species represent 94.4 percent of the birds counted at Hawk Mountain. In contrast, three *Falco* species form only 1.09 percent of the flights. Equally interesting, three species combined (two *Buteo* and one *Accipiter*) represent 90.43 percent of the flights: Broad-winged Hawks (41.86 percent), Red-tailed Hawks (24.96 percent) and Sharp-shinned Hawks (23.61 percent).

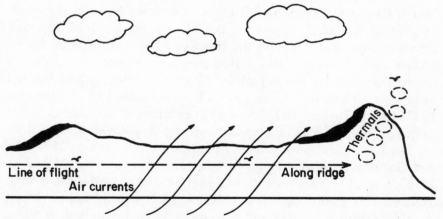

Line of flight   Along ridge
    Air currents   Thermals

FIGURE 19   A mountain ridge showing surface winds being deflected upward after striking the sides of the mountain. Migrating hawks soar on these updrafts but also may use thermals as shown at the right. Modified and redrawn after Simpson (1954: 18).

The flight styles of the two buteos are somewhat different. Red-tails are extremely dependent upon the use of strong updrafts, whereas Broad-wings can use updrafts, thermals, or combinations of both.

The Sharp-shinned Hawk, appears to use updrafts along ridges readily but is not necessarily dependent upon them. Allen and Peterson (1936) and Stone (1937), for example, report that the Sharp-shin is the most abundant species in the autumn hawk flights seen at Cape May Point, New Jersey. Cooper's Hawks appeared next in abundance, followed by Sparrow Hawks. Combining the various species into appropriate genera, we find *Accipiter* species are most abundant, and *Falco* species are ranked second at Cape May Point. However, a reduction in the numbers of Sharp-shins seen at Cape May Point may have occurred in recent years although good systematic data are difficult to obtain for this spot. Moreover, the extraordinary Sparrow Hawk flight reported by Choate (1972), coupled with the number of birds which necessarily may be overlooked during activities connected with raptor banding at the Cape (Clark, 1972), may have created a bias in recent Cape May Point hawk counts. Nonetheless, Choate and Tilly (1973) present the best available hawk counts for the 1970 season at Cape May Point.

The degree of dependence which Bald Eagles (*Haliaeetus*), Marsh Hawks (*Circus*), and Ospreys (*Pandion*) have on mountain updrafts is less obvious as each is well represented both in ridge (Broun, 1949a; Heintzelman and MacClay, 1972) and coastal flights (Allen and Peterson, 1936). However, the apparent increase in Ospreys seen along the mountain ridges is perplexing and has led to several speculative hypotheses (Heintzelman, 1970c; Peterson, 1966: 10–11; Taylor, 1971). No less fascinating is the Golden Eagle (*Aquila*). Golden Eagles from the remnant Appalachian population are famous attractions at Hawk Mountain (Broun, 1949a: 187–195) and Bake Oven Knob, Pennsylvania (Heintzelman and MacClay, 1972), but seldom are seen during migration at coastal stations such as Cape May Point, New Jersey. These birds usually appear along the ridges during October and November on cold, windy days when excellent updrafts are being generated (e.g., see Heintzelman, 1972d: 19–20).

Why Bald Eagles appear along the mountain ridges and along

the coastal flyway whereas Golden Eagles rarely appear on the coast during migration is not clear. Ecological niche requirements may be the answer. Bald Eagles are fish eaters and thus are normally associated with aquatic ecosystems whereas eastern Golden Eagles are birds of mountainous areas (Spofford, 1971). Hence one might expect that Golden Eagles would tend to remain inland and migrate along mountain ridges rather than use an ecologically unfamiliar coastal route. Nonetheless, Golden Eagles do appear along coastal areas during winter.

14

Hawk Migrations
and Thermals

The use of thermals by migrating Broad-winged Hawks is of major importance when considering autumn hawk migrations. It is essential to the success of these birds reaching their winter range in Middle and South America. Although the Swainson's Hawk, a midwestern and western species, also uses thermals extensively during migration, and frequently combines with Broad-wings in forming mixed flocks in Central America (Skutch, 1945), only the Broad-wing is important as an eastern North American thermal-dependent species. In considering Broad-wings, and their use of thermals during migration, one should first examine thermal formation, thermal structure, and then details of their use.

THERMAL FORMATION

Foster (1955), explaining how thermals are formed, points out that the radiation of the sun is the main source of heat and that, as the sun sets, a gradual diminishing of the heat in the air occurs. The heat is either radiated into space, to be reflected back to the earth or sea by clouds, or conducted or convected into the surrounding atmosphere. With the loss of heat, the air becomes

The River of Rocks at Hawk Mountain, Pa. The different heat-reflecting capacities of the boulders and surrounding forest, spread over a distance of about one-half mile, is adequate to produce thermal bubbles. Photo by Donald S. Heintzelman.

increasingly stable because hot and cold air mix and diffuse into a generally homogeneous mixture. Immediately before sunrise the atmosphere has reached its most homogeneous aspect. Measurements of air temperature at that time would demonstrate an even fall of temperature with height or the presence of an inversion. There are few thermals, and the atmosphere is quite unsuitable for soaring.

However, when the sun strikes the atmosphere, radiation begins to heat the air. If the atmosphere has a homogeneous composition of air and water vapor, the radiation causes an even rise in temperature. No air currents will be formed because there will be no change in pressure. Therefore, for thermal activity to start, there must exist a discontinuity in the rate at which the air is heated. This discontinuity is created when the sun irradiates land surface features such as rocks, sand, gravel, quarries, and towns—features which are of different thermal capacity from adjacent areas. Since

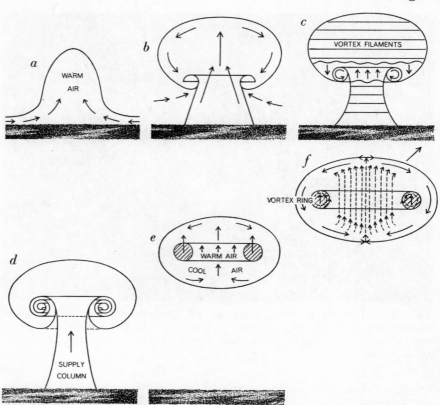

FIGURE 20 Mechanics of thermal bubble formation. Warm air on the ground
rises (*a*) and assumes a bubble-like shape (*b*). Turbulence at the surface of
the bubble forms twisting filaments that rapidly coil into a toroidal vortex ring
similar to a ring of smoke (*c*, *d*). The warm air mass is buoyant and is pinched
off from the ground by cool air flowing inward (*e*). The thermal bubble or
shell then floats away (*f*). The bubble then entrains some cool air from the
outside. This air circulates upward through the center of the vortex ring then
downward around the outside of the ring. Reprinted from "The Soaring Flight
of Birds" by Clarence D. Cone, Jr. Copyright © 1962 by Scientific American,
Inc. All rights reserved.

ground temperature rises more rapidly than air temperature, the
ground provides radiating surfaces which are in contact with the
air. This results in localized increases in air temperature. Eastwood
(1967: 184) points out that large differences in temperature are
not required for thermal production—fractions of a degree centigrade over horizontal distances of a half a mile are sufficient. At
Hawk Mountain, Pennsylvania, for example, Nagy (1973: 19) points

out that the River of Rocks—a mile-long boulder field below the Hawk Mountain lookouts—is an excellent thermal producer.

Foster (1955) continues, explaining that air which is sufficiently hot rises and forms a layer of varying depth which grows in size until it breaks away from the thermal source as a large bubble. This rises into the atmosphere and loses its heat by radiation and intermixing or it reaches a layer of warmer air again.

In many instances thermals can be observed. Since water vapor is present in the air within thermals, the ascending bubble eventually reaches a height where the dew point is reached and the water vapor appears as visible particles in the form of a cumulus cloud. These clouds begin forming from an hour and a half to three hours after sunrise and continue forming, in larger size, as the day progresses. When their development is unrestricted, they build into cumulo/nimbus thunderclouds of enormous proportions and altitudes and contain vast ascending currents.

Foster (1955) also explains that observations show that cumulus clouds generally emanate from one place because thermals tend to develop and rise from one area of ground and then are carried downwind. Thus long lanes of cumulus clouds are often observed with parallel spaces of clear sky moving in the direction of the wind. Cumulus clouds, which are fine weather clouds, grow and disappear rapidly, the average life being about twenty minutes.

The rate at which cumulus clouds grow and decay is an excellent indication of the presence of thermals. Sailplane pilots, for example, use this fact to achieve lift by flying from one cumulus cloud to another, gaining height by circling beneath a growing cumulus and losing it again on the passage to the next thermal. Similarly, Broad-winged Hawks apparently carry out their migrations with relatively little expenditure of energy by using the same technique.

For a more detailed technical explanation of thermal formation, the reader is referred to the works of Cone (1961, 1962a, 1962b).

THERMAL STRUCTURE

There are two important theories of thermal structure: (1) vortex rings within bubbles, and (2) continuous columns of rising warm air.

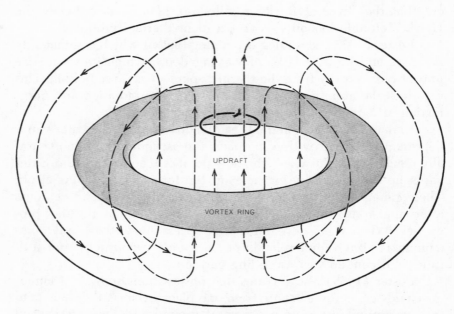

FIGURE 21 A thermal bubble or shell contains a vortex ring of warm air around which a current of cooler air circulates in a streamlined pattern (broken lines). A soaring hawk flies in a circle with the radius giving the hawk an aerodynamic sinking velocity equal to the updraft velocity. The bird thus is held in an equilibrium position inside the thermal bubble which, in turn, rises aloft and can be subject to wind drift. Reprinted from "The Soaring Flight of Birds" by Clarence D. Cone, Jr. Copyright © 1962 by Scientific American, Inc. All rights reserved.

VORTEX RINGS WITHIN BUBBLES

The most widely accepted explanation of thermal structure views the air mass as an invisible bubble, containing a vortex ring of warm air, rising into the atmosphere from its point of origin. Observations of kettles of Broad-winged Hawks milling around in a restricted mass with the general shape of a bubble tend to support this theory. Cone (1961, 1962a, 1962b) provides the most detailed explanation and drawings of thermal structure.

COLUMNS OF WARM AIR

An older theory of thermal formation and structure follows Foster's (1955) explanation of formation but describes the air mass as a continuous column of warm air rising from its source to the dew point, where cumulus cloud development signals the point of decay. This continuous column theory is not accepted currently although it appears in some of the older hawk migration literature.

RAPTOR USE OF THERMALS

When considering thermal use by migrating diurnal birds of prey in eastern North America, at least three related factors may be important: (1) raptor morphology, (2) seasonal aspects of hawk migrations, and (3) behavior.

RAPTOR MORPHOLOGY

Morphologically, the Broad-winged Hawk appears to be well-designed to use thermals, and its observable behavior supports this. Its aspect-ratio is 2.28±0.02 (a low value), and its wing loading is 0.37. In comparison, the Red-tailed Hawk (which seldom uses thermals) has a wing loading ranging from 0.46 to 0.56. Ospreys, which occasionally use thermals, have an aspect-ratio of 3.00 and a wing loading of 0.51 to 0.65—values considerably higher than those for the Broad-wing. Likewise, the Peregrine Falcon, which does relatively little thermal soaring, also has an aspect-ratio of 3.06 and a wing loading ranging from 0.62 to 0.91.

However, the Sharp-shinned Hawk has an aspect-ratio of 2.18 and a wing loading ranging from 0.22 to 0.31. These are values lower than those for the Broad-winged Hawk. Nonetheless, Sharp-shins are not particularly dependent upon thermals. They use them only occasionally and never to the extent that Broad-wings do. In fact, most hawks, eagles, and falcons use thermals occasionally.

SEASONAL CONSIDERATIONS

The seasonal restrictions on autumn hawk migrations in eastern North America are crucial when considering raptor use of thermals. Each species has a definite range of dates between which migrations occur. More important, a fairly well-defined peak migratory period exists between the extremes of the dates. It is during this peak period that the bulk of the individuals are seen. Broun (1949a), and

Haugh (1972), for example, charted the seasonal aspects of autumn hawk migrations at Hawk Mountain, and Heintzelman (1969b, 1972d) charted similar data for Bake Oven Knob, Pennsylvania. Similarly, Haugh (1972: 16–17) charted seasonal data for each species appearing at Hawk Cliff, Ontario. Mueller and Berger (1961) provided similar data for Cedar Grove, Wisconsin. Collectively, this information adequately describes the seasonal aspect of autumn hawk migrations in the East.

It is general knowledge, therefore, that mid-September is the peak migratory period for the Broad-wing in the northeast. Later, in October and November, when colder air temperatures occur, other nonthermal-dependent hawks replace Broad-wings as the most abundantly seen migrants. But within the seasonal framework of the Broad-wing's autumn migration period, mid-September presumably provides the most favorable thermal activity for this species in the northeast. Further southward peak Broad-wing flights occur a little later. In Tennessee, for example, Broad-wing flights often are seen about a week later than in Pennsylvania, and the largest flights do not occur in Central America until October.

However, local weather conditions often influence Broad-wing flights greatly. It is not uncommon for lookouts located only a few miles apart to record peak flights on entirely different days. However, geography also plays an important role in determining where migrating hawks will appear.

BEHAVIOR IN THERMALS

No discussion of autumn hawk migrations is complete without a description of the behavior of Broad-winged Hawks in thermals. The sight of hundreds of these birds confined within a thermal bubble is one of the most spectacular phenomena in nature. It resembles a boiling mass of hawks milling within the confines of an invisible balloon. As the kettle rises into the atmosphere, there is no apparent organization among the birds. However, as the thermal approaches the dew point, and gradually cools and expands, energy within the thermal escapes, and the hawks begin gliding downward, either singly or a few at a time, eventually to enter another developing thermal. The best time to count Broad-wings is when they

Upper left, a Broad-winged Hawk circling in a thermal. Photo by Donald S. Heintzelman. *Upper right,* Broad-winged Hawks gliding out of a thermal. The birds sometimes glide for several miles before entering another thermal. Photo by Donald S. Heintzelman. *Below,* a kettle of Broad-winged Hawks inside a thermal. Photo by Donald S. Heintzelman.

leave a thermal and stream across the sky in bomberlike formation as they seek another thermal.

It is by this process of riding thermals aloft, then gliding downward to new ones, that Broad-winged Hawks successfully carry out long-distance migrations and eventually reach their winter range in Central and South America. Although updrafts along mountains and lake shorelines may be used for varying distances, they are much less important to Broad-wings than to many other species of hawks, eagles, and falcons. Foster (1955) provides an excellent sketch of the general mechanics of Broad-wing thermal use.

One of the earliest descriptions of the behavior of migrating Broad-winged Hawks using thermals was published by von Lengerke (1908): "I was on the top of a mountain near Stag Lake, Sussex County, N.J., about 1200 feet above sea level, from where I had an unobstructed view for miles of country all around me. My object was to observe the migration of hawks, and I was armed with a Hensoldt Binocular eight power glass. The day was clear, and at one time in the forenoon, several thousand hawks, Broadwings mostly, were in view. They came from a northeasterly direc-

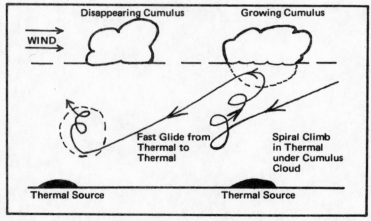

FIGURE 22 The mechanics of thermal use by migrating Broad-winged and Swainson's Hawks (and other species occasionally) involves the birds entering a thermal bubble, riding it aloft until it decays, then rapidly gliding downward and entering a new thermal to repeat the process. "Kettles" are groups of hawks inside a thermal bubble. Reprinted from Heintzelman (1972d) after being redrawn from Foster (1955).

tion which would take them directly to the Shawangunk Mountain, Ellenville, and Lake Minnewaska, N.Y., sixty miles northeast from my place, where a similar flight was observed by Mr. Barbour and Mr. Kirk Monroe. A constant stream of birds, very high up, could be seen for a long while, and they were going in the direction of the Delaware Water Gap. Over the valley to the southwest of me, the birds seemed to collect into an immense flock, while hundreds, if not thousands of birds were gyrating around and around, describing smaller and larger circles in the air, in heights of from, I would judge, 600 to 2,000 feet above the earth. Most birds were Broad-wings. There were, however, other hawks such as Red-tails and Red-shoulders among them, while the 'Accipiter' genus was represented by some Cooper's Hawks and more Sharp-shinned, which, however, were mostly flying lower and took no part in the general evolution."

Occasionally other species of hawks also use thermal soaring flight. Unlike Broad-wings, however, these usually are individual birds, although sometimes half a dozen or so may use a thermal together.

15

Daily Rhythms and
Noon Lulls in
Autumn Hawk Flights

When considering the behavior of migrating hawks, eagles, and falcons in eastern North America in autumn, two facets should be examined: (1) trends in the daily rhythm patterns of these birds, and (2) noon lulls occurring in some flights of certain species.

DAILY RHYTHM TRENDS

Observers at various hawk lookouts in the East long have informally discussed the daily rhythm patterns of migrating diurnal birds of prey. However, Mueller and Berger (1973) correctly point out that relatively little information has been published on the subject. At Cape May Point, New Jersey, Allen and Peterson (1936) report that most Sharp-shins appear in the morning whereas Pigeon Hawks are most apt to appear in the afternoon. At Hawk Mountain, Pennsylvania, Broun (1939, 1949a) reports that the majority of Sharp-shins seen during September occur during the afternoon. Another long-standing idea, held by some observers at the Pennsylvania lookouts, suggests that eagles are likely to appear late in the afternoon (Broun, 1949a: 164). Since many of these ideas are based upon subjective impressions, it seems worthwhile to examine some of them in more detail.

To examine trends in the daily rhythm patterns of migratory behavior of hawks, eagles, and falcons at Cedar Grove, Wisconsin, Mueller and Berger (1973) quantified data covering a four-year period and determined that "The autumnal migration of accipiters peaks in the early morning, that of buteos in late morning, and that of falcons in early afternoon. Harriers have an early peak, and Ospreys show no peak." However, Broad-winged Hawks, unlike other buteos, had peaks appearing in midmorning and late afternoon.

Cedar Grove is located along the west shore of Lake Michigan. Since conditions and behavior patterns of migrant hawks at this site may differ from those along a mountain flyway (Mueller and Berger, 1973), I have quantified autumn data from Bake Oven Knob, Pennsylvania, for the period 1968 through 1971, following methods used by Mueller and Berger (1973). Data were pooled

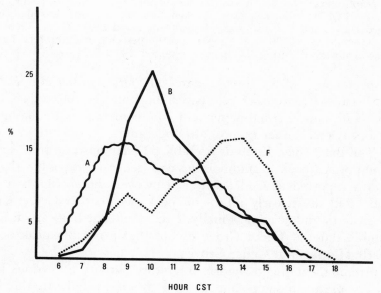

HOUR CST

FIGURE 23 The daily rhythm of migrating hawks at Cedar Grove, Wis. The graph shows the hourly distribution of migrant *Accipiter* species (wavy line, n = 4,602), *Buteo* species (solid line, n = 2,317) less the Broad-winged Hawk, and *Falco* species (dotted line, n = 987). The value for each hour interval (e.g., 0700–0759) represents the percentage of the total number of hawks observed during the four year period 1958–1961. Reprinted from Mueller and Berger (1973).

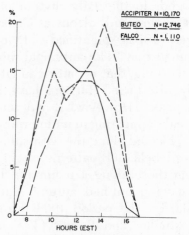

FIGURE 24   The daily rhythm of migrating hawks at Bake Oven Knob, Pa.
The graph shows the hourly distribution of migrant *Accipiter* species (solid
line, n = 10,170), *Buteo* species (broken line, n = 12,746) less the Broad-
winged Hawk, and *Falco* species (dashed line, n = 1,110). The value of each
hour interval (e.g., 0700–0759) represents the percentage of the total number
of hawks observed during the four year period 1968–1971.

for species in each of three genera—*Accipiter*, *Buteo* and *Falco*—
with Broad-winged Hawk counts deleted from the *Buteo* data since
Broad-wings are thermal-dependent and, compared with the migratory behavior of other buteos, are abberant.

The Bake Oven Knob data (table 53) demonstrate that *Accipiter* migrations peak in midmorning and generally resemble rhythm
patterns reported at Cedar Grove. However, *Buteo* flights at the
Knob build to an early afternoon peak, whereas at Cedar Grove
they peak in midmorning. Finally, *Falco* flights at Bake Oven Knob
resemble those at Cedar Grove except that a notable peak occurs
in early afternoon at Cedar Grove.

Although graphs of hourly migratory activity at various lookouts are useful in suggesting general trends in daily rhythm patterns of hawks, many flights deviate markedly from these trends.

NOON LULLS: BACKGROUND INFORMATION

A particularly curious feature of autumn hawk migrations is
the lull which often develops in some hawk flights during the mid-

dle of the day. These noon lulls, as they are called, can vary by an hour or more on either side of noon, and some flights apparently do not contain them. However, it is perhaps notable that flights without lulls usually occur on days when heavy cloud cover prevails. Nonetheless, noon lulls are well known to hawk watchers in Pennsylvania. Broun (1949a: 152), for example, mentions their regular occurrence during September and October hawk flights at Hawk Mountain. Heintzelman and Armentano (1964) report similar lulls in many of the Bake Oven Knob flights. Moreover, data from various other lookouts in eastern Pennsylvania and New Jersey also show noon lulls in some flights. In fact, the phenomenon frequently occurs at many northeastern lookouts.

What, then, are noon lulls? Do they actually exist? Why should they occur? What causes them? And, how do they influence the accuracy of hawk counts and possibly the routes used by migrant hawks?

This discussion is limited to an examination of the noon lull phenomenon in eastern Pennsylvania. However, there is no reason to believe that different circumstances cause lulls in hawk flights elsewhere.

Species for which noon lulls in some daily flights are especially easy to detect are Sharp-shinned, Red-tailed, and Broad-winged Hawks. These are the three most abundant raptors in the autumn hawk flights in eastern Pennsylvania. Although lulls also may occur in flights of other species, they occur in much smaller numbers, and it is difficult to detect lulls in their flights. Nonetheless, an examination of certain of these less abundant species, notably eagles, may be worthwhile.

DESCRIPTIONS OF NOON LULLS

Before attempting to explain the cause or causes responsible for creating noon lulls in hawk flights, it first seems desirable to describe some typical examples. The graphs for Sharp-shins, Red-tails, and Broad-wings represent data for single days, whereas the eagle graphs represent pooled data for all available years (1961 through 1972) grouped into frequency distributions based upon one-hour intervals. It was necessary to treat the eagle data in this manner since too few birds appear on any single day to permit any patterns to be detected.

SHARP-SHINNED HAWK

I have selected four examples of Sharp-shinned Hawk flights to illustrate various patterns in noon lulls for this species. Lulls

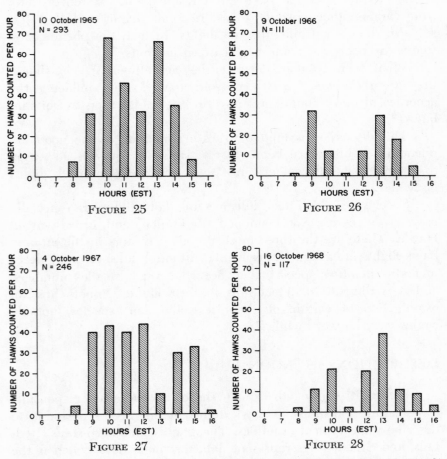

FIGURES 25–28 The hourly distribution of selected Sharp-shinned Hawk flights at Bake Oven Knob, Pa. Noon lulls are more evident in some flights than in others.

were more pronounced in some flights than in others. Nonetheless, they occurred in each of the flights illustrated.

A curious sidelight on the noon lull phenomenon in Sharp-shin flights at Hawk Mountain, Pennsylvania, *during September,* is that the bulk of these birds *"always"* appear during the afternoon whereas at Cape May Point, New Jersey, the birds fly largely during the morning (Broun, 1949a: 157). At Cedar Grove, Wisconsin, Sharp-shin flights also peak in the morning (Mueller and Berger, 1973).

Since a mountain flyway might have a different influence upon migration patterns of hawks as compared with coastal and lake-shore sites, I have evaluated the phenomenon for Bake Oven Knob, Pennsylvania. September data for the eight-year period 1964 to 1971 were extracted from my notebooks. Each year's data were grouped into the number of birds seen during the morning and the number seen during the afternoon (table 54). Considered collectively, slightly more Sharp-shins were seen in the afternoon in September than in the morning. However, during some years more birds appeared during the morning, and during other years more appeared during the afternoon. But when all *Accipiter* data for four *complete* seasons were pooled, and the hourly numbers of birds determined, and expressed as a percentage of the total, accipiters peaked during the morning.

RED-TAILED HAWK

The Red-tailed Hawk is one of the largest soaring hawks passing Bake Oven Knob during autumn. The bulk of the flights occur in late October and early November (Heintzelman, 1969: 16). Broun (1949a: 151, 158–159) reports a similar peak period at Hawk Mountain.

Red-tails are primarily soaring birds, and they make extensive use of updrafts which occur along the Kittatinny Ridge and other mountains. Occasionally a few Red-tails also make limited use of thermals (Broun, 1949a: 158–159), but never to the extent that Broad-winged Hawks do. One never observes large flocks of Red-tails in kettles although five or six birds might be seen together in a thermal on rare occasions.

Despite their limited use of thermals, Red-tails often exhibit noon lulls in their flights—especially on days with blue skies and limited cloud cover. The graphs (figs. 29–32) illustrate four typical Bake Oven Knob Red-tail flights with noon lulls.

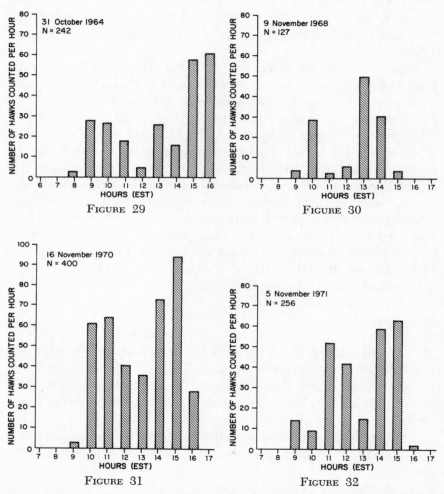

FIGURE 29

FIGURE 30

FIGURE 31

FIGURE 32

FIGURES 29–32   The hourly distribution of selected Red-tailed Hawk flights at Bake Oven Knob, Pa. Noon lulls are more evident in some flights than in others.

BROAD-WINGED HAWK

Noon lulls frequently appear in Broad-winged Hawk flights, apparently as a result of the dependence of these birds upon thermals. For example, on cloudless days (or days with limited numbers of clouds) in mid-September, observers commonly see Broad-wings rising to extreme altitudes, massed in flocks or kettles inside thermals. Eventually they are so high that they appear as pepper specks against the zenith. It is nearly impossible to observe and count hawks under such circumstances, and observers frequently strain their eyes to avoid missing birds. Nonetheless, large numbers of Broad-wings probably are overlooked under such conditions. The result is a so-called lull in the day's flight. Hence noon lulls in Broad-winged Hawk flights in particular probably exert a great influence upon the accuracy of Broad-wing counts. Some typical examples of noon lulls in Broad-wing flights at Bake Oven Knob, Pennsylvania, are illustrated (figs. 33–39).

GOLDEN EAGLE AND BALD EAGLE

The number of Golden and Bald Eagles which daily pass Bake Oven Knob in autumn is relatively small. Hence it is impossible to detect noon lulls in eagle flights by a direct examination of the daily data. However, by pooling and tabulating into hourly distributions the times of all eagle records from the Knob for the period 1961 through 1972, it is possible to look for a noon-lull trend in eagle flights. One can also test the validity of the idea that eagles usually appear late in the afternoon.

The Golden Eagle data (table 55) reveal two peaks, one in midmorning and another in early afternoon. Thirty-two percent of the birds are seen prior to noon and 68 percent occur during the afternoon. A slight lull appears between 1100–1200, but no large late afternoon flight is evident.

In contrast, Bald Eagle flights (table 55) peak between 1100 and 1200 and show no noon lull. Thirty-eight percent of the birds appear during the morning, and 62 percent are seen in the afternoon, chiefly early afternoon. Thus, despite the late morning peak,

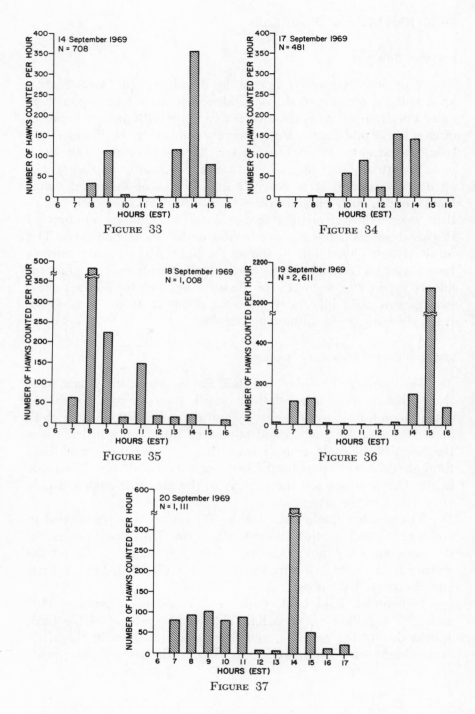

FIGURE 33

FIGURE 34

FIGURE 35

FIGURE 36

FIGURE 37

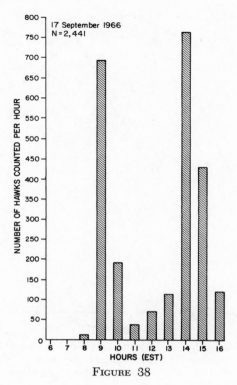

FIGURE 38

FIGURES 33–38 The hourly distribution of selected Broad-winged Hawk flights at Bake Oven Knob, Pa. Conspicuous noon lulls are evident in each of these flights.

the Bake Oven Knob data partly confirm Broun's (1949a: 164) statement that most Bald Eagles appear in the afternoon (at Hawk Mountain). However, the Bake Oven Knob data do not confirm the statement that eagles migrate very late in the day at Hawk Mountain. Only a relatively few birds do so at the Knob.

HARRIERS AND OSPREYS

Harriers or Marsh Hawks also occur in relatively limited numbers at Bake Oven Knob, thus making it difficult to detect lulls in daily flights. Hence data for four seasons (1968–1971) were pooled

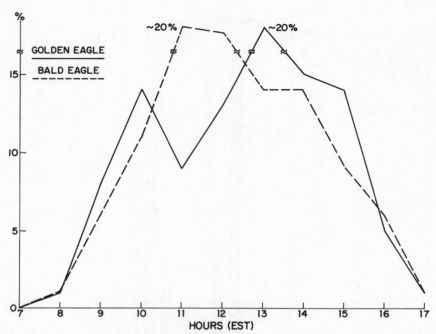

FIGURE 39   The daily rhythm of migrating Golden and Bald Eagles at Bake Oven Knob, Pa. The value for each hour interval (e.g., 0700–0759) represents the percentage of the total number of each species observed during the period 1961–1972.

and arranged in one-hour intervals (table 53). This shows a peak in mid- to late morning but no evidence of a noon lull. Mueller and Berger (1973) report a similar, but somewhat earlier, morning peak for harriers at Cedar Grove, Wisconsin.

At Cedar Grove, Mueller and Berger (1973) also report that Ospreys occur ". . . with about equal frequency throughout the hours between 07:00 and 16:59" although their data, in fact, show some degree of variation throughout the day. In contrast, at Bake Oven Knob, Pennsylvania, a markedly different daily rhythm in Osprey flights occurs as revealed by a pooling and hourly distribution of four season's data (table 53). Here the flight tends to build to an early afternoon peak, followed by a very slight lull, then another peak in midafternoon.

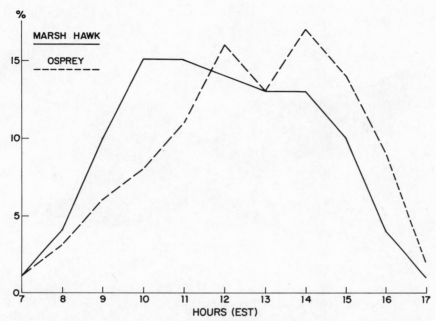

FIGURE 40 The daily rhythm of migrating Marsh Hawks and Ospreys at Bake Oven Knob, Pa. The value for each hour interval (e.g., 0700–0759) represents the percentage of the total number of each species observed during the period 1968–1971.

POSSIBLE CAUSES OF NOON LULLS

Much additional field study is needed to fully understand noon-lull phenomena. Radar equipment might be especially useful in tracking the movements of birds. Nonetheless, it seems useful to advance various possible causes of noon lulls to serve as a basis of future investigations.

LANDING AND FEEDING

This is the oldest and most common explanation. Broun (1949a: 152), for example, states that hawks presumably stop to feed and rest during the noon lull although he seldom saw hawks doing so at Hawk Mountain. Similarly, at Bake Oven Knob in 1957,

A Red-tailed Hawk hanging motionless in midair over the Kittatinny Ridge. The bird is looking for prey on the forested slopes below. Photo by Fred Tilly.

and between 1961 and 1973, 799 days of observation resulted in 145,375 hawks being counted. Very rarely, probably less than a dozen times, have I observed hawks landing during lulls. However, I have on occasion watched Broad-winged Hawks and Sparrow Hawks capturing butterflies and other insects in midair, then carrying and eating the prey while remaining in flight. Similarly, I have frequently observed hunting Red-tailed Hawks "hanging" motionless in midair, intently watching the ground for prey. Sometimes they plummet to the earth, presumably to capture prey, but only rarely do these birds appear to land. Nonetheless, if goodly numbers of Red-tails were distributed along the Kittatinny Ridge and engaged in hunting activities, a lull might occur in the flight. This idea cannot be completely dismissed. For example, on 24 October 1964 about one-half of the Red-tailed Hawks observed passing Bake Oven Knob after 1400 had full crops, indicating that the birds had been hunting successfully. On 6 November 1966, I again saw a few Red-tails with full crops during the middle of the afternoon, and on numerous other occasions birds with bulging crops were seen at Bake Oven Knob and other lookouts.

Similarly Ospreys occasionally are observed passing the Pennsylvania hawk lookouts carrying a fish in their talons, again demonstrating successful Osprey hunting activity. The fish, incidentally, always is positioned with its head pointing forward. And once, at Tott's Gap, Pennsylvania, I observed a migrating Sharp-shinned Hawk carrying an unidentified prey item in its talons. Hence, for some species, feeding activity might be a minor contributing factor in the development of noon lulls in hawk flights. Along the north

Migrating Red-tailed Hawks with full crops sometimes are seen passing lookouts. This indicates that they have fed recently. The birds halt their migrations briefly while hunting and feeding. Photo by Donald S. Heintzelman.

shore of Lake Erie in autumn, Haugh (1972: 6) also observed groups of migrating Sparrow Hawks pausing for five to ten minutes to feed upon grasshoppers.

Nonetheless, I have never seen a migrating hawk land to eat prey captured during a noon lull. Hence I doubt if landing and feeding is an important factor in creating noon lulls in hawk flights.

INTERSPECIFIC BEHAVIOR

At Bake Oven Knob, Hawk Mountain, and most northeastern hawk lookouts, interspecific displays between various raptors are seen occasionally. Sharp-shinned and Cooper's Hawks, for example, sometimes harass other hawks for brief periods of time. Occasionally, however, these interspecific displays are more intensive or violent, or continue for more prolonged time periods.

At Hawk Mountain, Pennsylvania, for example, Broun (1949a: 194) observed a dramatic encounter between a Golden Eagle and a Red-shouldered Hawk in which the eagle captured the hawk, after being harassed by the smaller bird, and plunged to the earth clutching the hawk in its talons. This extraordinary incident is one of the few examples of a migrating diurnal raptor capturing prey and landing with it. Moreover, Broun states that this was the only instance he had ever seen of one raptor capturing another.

Over the years at Bake Oven Knob, Pennsylvania, I have observed, on various occasions, interspecific behavior of more than momentary passing or interest. One of the most intense and prolonged examples, between a Sharp-shinned Hawk and a Sparrow Hawk, occurred on 12 September 1967 (Heintzelman, 1970f):

"At 1420 hours, I briefly left the South Lookout to reset a string of mammal traps running along the north slope of the Kittatinny Ridge at the Knob.

"Suddenly a sharp 'cackling' vocalization attracted my attention. I looked skyward just as an immature Sharp-shinned Hawk (*Accipiter striatus*) appeared low overhead and darted past me. A male Sparrow Hawk (*Falco sparverius*) was in close pursuit of the Sharpie. Three or four times the two birds circled above the spot where I stood. As the display continued, the role of attacker alternated.

Interspecific displays between hawks sometimes occur and may cause a brief disruption in their migrations. Here a Cooper's Hawk (right) has been harassing a Broad-winged Hawk (left). Photo by Donald S. Heintzelman.

"Suddenly the Sharpie again darted past me and landed in an oak tree some 20 feet away. Seconds later it fled skyward with the falcon still in pursuit. Again the position of attacker alternated as the two birds circled, dove, and shot skyward. At times each bird made nearly direct strikes at the other but I could discern no actual physical contact. The birds appeared to be nearly evenly matched in speed and agility. However, the *Accipiter* was definitely more maneuverable when darting between the trees and shrubs which partially covered the rock outcropping where I stood.

"About five minutes after my initial awareness of the birds, a second male Sparrow Hawk joined in the display. Each falcon took turns in the attack. As one completed a stoop at the *Accipiter*, the other began its stoop. Unfortunately, after I had observed the display for at least eight minutes, all three birds disappeared from view as they flew southwestward along the Kittatinny Ridge in the general direction taken by migrant hawks that day. During the entire period of my observations, only the Sharp-shinned Hawk made vocalizations."

On another occasion I watched a Peregrine Falcon cruise back and forth along the ridge in the vicinity of Bake Oven Knob, an activity which clearly created a disruption in its migration. If a sizeable number of migrating hawks regularly engaged in interspecific behavior for prolonged periods, lulls might develop in their flights. However, the examples cited, and similar examples, are seen only occasionally. Hence interspecific behavior seems in no way to be responsible for creating noon lulls in hawk flights.

LOW-ALTITUDE PARALLEL-TO-MOUNTAIN FLIGHTS

Another explanation is that hawks vary the distance from a mountain as they fly parallel to it. Thus, during lulls, hawks presumably would fly at a greater distance from a mountain than during other periods. Hence the birds would be more difficult to detect during the lull. This occurs at times, as observers at Bake Oven Knob sometimes barely are able to see hawks flying low, but parallel to, the mountain as they cross a valley immediately north of the Kittatinny Ridge. Since this situation does not occur with great frequency, I doubt if it is the primary cause responsible for creating most noon lulls, however.

HIGH-ALTITUDE PARALLEL-TO-MOUNTAIN FLIGHTS

Another version of the previous explanation suggests that hawks fly at extremely high altitudes, following a course parallel to the mountain but far out over the valley north of the Kittatinny Ridge. This hypothesis also requires that birds alter both the altitude at which they are flying and the distance from the mountain, if noon lulls are to occur due to its influence.

There are many occasions when hawks fly past Bake Oven Knob very high, and far out over the north valley, but the birds nonetheless are visible through binoculars and/or telescopes. For example, this type situation produced a skewed distribution for the

Red-tailed Hawk fligʰ.t for 26 October 1968. Because of the distances of the hawks from the lookout, and poor light conditions, a
20× telescope was used to identify the 161 Red-tails which passed
the Knob. In any event, my experience at Bake Oven Knob does
not lead me to believe that high-altitude parallel-to-mountain
flights normally are responsible for causing noon lulls.

DIAGONAL BROAD-FRONT FLIGHTS

Still another explanation suggests that hawks are flying north
of the Kittatinny Ridge but slowly are being drifted by wind as
they move over a broad front. Thus, as hawks diagonally approach
a mountain, they may be counted at one site but not at another.
For example, on 18 September 1965, Alan and Paul Grout were
stationed on Bear Rocks less than two miles southwest of Bake
Oven Knob. They noticed 335 Broad-winged Hawks flying toward
them from a northerly direction. These birds were not seen upridge
at Bake Oven Knob. Apparently the hawks were migrating over a
broad front. Upon reaching the mountain, they then followed it as
a diversion-line * for an unknown distance. A similar example
occurred on 14 September 1963, between Bake Oven Knob and
Hawk Mountain (Heintzelman and Armentano, 1964: 8–10). Nagy
(1970: 9) also reports that a flight of Broad-wings at Hawk Mountain approached, and crossed, the ridge at right angles under conditions of strong northerly winds. Kettles of hawks were seen in all
directions at the limit of his vision. Evidently this was also a broad-
front migration.

However, even broad-front migrations fail to explain noon
lulls completely. A wave-like migratory movement would be required, spread over a broad front, for lulls to occur. Hopkins and
Mersereau (1971: 3) suggested a wave-like migratory movement
for migrating Broad-winged Hawks in Connecticut, but their data
are fragmentary and not convincing. Hence diagonal broad-front
flights also seem to be unlikely primary causes of noon lulls.

* A topographic feature, usually long, which induces migrating birds to alter their
direction of flight, and diverts them in a new direction. The new direction of flight
then is followed for varying distances.

HIGH-ALTITUDE FLIGHTS

In my view, the most likely explanation for the cause of noon lulls in hawk flights, especially Broad-winged Hawk flights, is the tendency for these birds to fly at increasingly higher altitudes during the warmest portion of the day. This is usually around noon or early afternoon. This idea is not entirely new. Stone (1937), for example, pointed out that many hawk flights at Cape May Point, New Jersey, occur at altitudes above the limit of man's vision. Similarly, at Bake Oven Knob, Pennsylvania, this behavior occurs most commonly when blue skies and few clouds prevail. Hawks then ascend to extreme altitudes, perhaps as high as 10,000 feet according to Murton and Wright (1968). However, Haugh (1972: 38) reported buteos using thermals at altitudes up to 5,000 feet, and Stearns (1948b) reported lower altitudes for Broad-wings in New Jersey. Nonetheless, I consider that most noon lulls, at least insofar as Broad-winged Hawks are concerned, are the direct result of high-altitude flights over a lookout and beyond man's vision. My evidence for this point of view follows.

Broad-winged Hawk

In considering noon lulls in Broad-wing flights, first review the data presented earlier regarding raptor morphology and anatomy, the relationship of these factors to bird flight, avian use of high-energy sources found within thermals, and the unusually high degree of dependence which Broad-winged Hawks have upon thermals when undertaking their migrations. Recall, too, that peak thermal activity usually develops around noon or early in the afternoon —generally the period when lulls occur in hawk flights, and generally the period when Broad-wings are observed as specks against the zenith.

To demonstrate the extreme altitudes which Broad-winged Hawks sometimes attain, some examples will be useful. Broun (1961: 8–9), for example, observed Broad-wings flying "exceedingly high and scarcely discernible as they moved through thin clouds" at Hawk Mountain, Pennsylvania, on 21 and 22 September 1960. Similarly, on 14 September 1965 at Bake Oven Knob, Pennsylvania, dense fog covered the summit of the Knob during the morning resulting in visibility of 100 feet or less. When the fog lifted around noon a Broad-wing flight began and continued until about 1700 hours. Between 1400 and 1500, I noted many Broad-wings flying at an extreme altitude and partly obscured by opaque mist from clouds. Some birds even disappeared into the opaqueness of the clouds, and it is possible that hawks may have been flying above clouds. Nonetheless, I counted 1,906 Broad-wings.

Another example occurred on 15 September 1967. Despite a poor Broad-wing flight at Bake Oven Knob, Hawk Mountain observers reported a good flight. For a week prior to the 15th, birds were moving over a broad migratory front. Few birds utilized the Kittatinny Ridge as a diversion-line. But at 1120 hours on 15 September a group of 81 Broad-wings appeared high over Bake Oven Knob. I watched them flying out of an opaque cloud layer which had developed thirty minutes earlier. This observation suggests that the hawks may have been flying above the clouds and dropped to a lower altitude over my lookout as thermal activity decreased. Presumably these birds would have remained beyond man's vision and would not have been counted had the opaque cloud cover not developed earlier.

On 19 September 1969 at Bake Oven Knob 2,611 Broad-winged Hawks were counted. This was the 1969 season's peak count for that site. During the morning of the 19th easterly winds at 5 to 20 miles per hour occurred. Only a few hundred hawks appeared against a background of blue skies with scattered cumulus clouds. Most of the birds passed over the valley far south of the Kittatinny Ridge. Moreover, we repeatedly watched Broad-wings riding thermals aloft earlier in the morning. During the next five hours few hawks were seen. By 1400 the blue skies changed to total opaque cloud cover, and thermal activity doubtless was curtailed drastically. Only 157 Broad-wings were counted between 1400 and 1500. Then, during the next hour, 2,099 Broad-wings glided overhead out of dense opaque clouds. As the hawks came into view they attempted to kettle using whatever weak thermals still existed south of the Kittatinny Ridge. Apparently thermal formation was nearly terminated, however, as seemingly endless formations of hawks glided lower and lower, past the lookout, until they finally landed on the wooded slopes of the ridge.

On 20 September 1969, between 0900 and 1100, I again observed Broad-wings gliding out of thermals and leaving the tops of cumulus clouds. Later in the afternoon, the hawks slowly lost altitude as opaque cloud cover developed. A marked noon lull occurred in this flight.

These and similar observations of Broad-winged Hawks flying into, through, above, or out of clouds are not unique to Bake Oven Knob. On 14 September 1926, Forbes and Forbes (1927) observed what apparently were Broad-wings using thermals over Mount Monadnock, New Hampshire. The birds gained great altitude and, to the astonishment of the observers, disappeared into clouds and reappeared again diving into the mist. "Finally the whole flight had spiraled upward into the cloud mass and was lost to view. Once, half a minute later, a few specks wheeled out toward us and for a moment could be dimly seen through the edge of the cloud. That was the last glimpse." The observations were made at 1150 hours. The observers estimated the height of the clouds at 7,000 feet.

The previous observations demonstrate that Broad-winged Hawks can ride within thermals to extremely high altitudes. Once high overhead, on days with blue skies, they are often unable to be seen unless opaque clouds provide a suitable background, and they can pass uncounted, thus creating an apparent lull in a day's flight. Indeed, it is likely that only the lowest-flying of these hawks can be seen under these conditions, and it is only these birds which constitute a day's count during the period of the noon lull. Hence a serious observer bias occurs in the count. But if thermal activity is reduced sharply as, for example, by rapid development of opaque cloud cover, more hawks will drop to lower altitudes. These birds then become more readily visible, thus creating an afternoon peak in the daily rhythm of the flight.

Clearly, Haugh (1972: 3) is in error in suggesting that clouds are generally unimportant to migrating hawks. Moreover, it is thermal formation rather than updraft formation which is affected by the development of an opaque layer of clouds. In the future, radar studies at major hawk concentration points may add a good deal of additional data to the currently available visual information on this subject. Presumably such electronic data could rapidly confirm or refute the views here presented.

Sharp-shinned Hawk

It is difficult to explain the cause of noon lulls in Sharp-shinned Hawk flights based upon thermal use, since this species seldom concentrates in thermals although individual birds sometimes use them briefly to gain altitude. In point of fact, Sharp-shins primarily are dependent upon updrafts along mountains or shorelines to aid them in their autumnal migrations.

How, then, can one explain the cause or causes responsible for creating noon lulls which sometimes occur in Sharp-shin flights? Stone (1937), commenting on flights at Cape May Point, New Jersey, suggested that the birds merely are riding updrafts to altitudes beyond man's limit of vision. Although this may not be a complete explanation for noon lulls in Sharp-shin flights, it seems to be the most likely cause at inland sites such as Bake Oven Knob, too.

On the other hand, some Sharp-shin flights do not produce lulls. On 6 October 1968, for example, 257 Sharp-shins passed Bake Oven Knob. These birds exhibited a daily rhythm suggesting a normal distribution. However, it is perhaps significant that opaque cloud cover occurred and the hawks were flying well within the limits of vision of the observers.

Red-tailed Hawk

The Red-tailed Hawk is the third most abundant species to migrate past Bake Oven Knob during autumn. Red-tails make extensive use of updrafts but only rarely use thermals during migration. Despite their use of updrafts along mountains rather than thermals, noon lulls sometimes occur in their flights. It appears that these lulls are caused by birds flying at extremely high altitudes, thus making it extremely difficult for observers to see the birds against a blue sky background. However, lulls also can occur on days when partial or even total opaque cloud cover prevails. On 2 November 1968, for example, a modest noon lull developed

despite about a 90 percent cloud cover at Bake Oven Knob. I feel
certain that the birds were flying too high, and far over the north
valley, to be seen.

Much additional study can be devoted to lulls in Red-tailed
Hawk flights until the phenomenon is adequately understood, however. Again, radar tracking might be very useful.

NOON LULLS AND HAWK COUNTS

Noon lulls can acquire considerable importance when considered in relation to the accuracy of hawk counts being made at a
concentration point such as Hawk Mountain or Bake Oven Knob.
If, during a lull, unknown numbers of birds pass uncounted, a bias
will occur in the day's count. If large numbers of birds were missed,
the bias will be serious. Moreover, if the percentage of uncounted
birds varies daily, and from lookout to lookout, as seems likely,
yearly hawk counts from a lookout will be incomplete reflections
of the actual number of hawks passing through the area. Hence
attempts to use autumn hawk counts from any single concentration
point can lead to highly inaccurate and misleading estimates of
raptor populations.

Part 5
Migration Routes,
Geography,
and Hawk Counts

16

Migration Routes
and Geography

It is impossible to discuss autumn hawk migrations without considering the influence of geography on migration routes. Hence this chapter examines the various migration routes, their more important geographic features, and the influence of these features upon migrating hawks.

MAJOR MIGRATION ROUTES

Three major hawk migration routes are known for eastern North America (Rusling, 1936). The Atlantic coastline is used by large numbers of accipiters and falcons, and smaller numbers of other species (Allen and Peterson, 1936; Ferguson and Ferguson, 1922; Rusling, 1936; Stone, 1937; Trowbridge, 1895). Inland, the Appalachian Mountains, notably the Kittatinny Ridge in New Jersey and Pennsylvania, form a second major route for soaring hawks, considerable numbers of accipiters, and some falcons and other species (Broun, 1949a; Frey, 1940; Heintzelman, 1969b; Tilly, 1972a, 1972b, 1973). A third route occurs along the shorelines of the Great Lakes (Pettingill, 1962), particularly at Cedar Grove, Wisconsin (Mueller and Berger, 1967) and at Duluth, Minnesota

(Hofslund, 1954, 1966). Other important Canadian concentration points along the Great Lakes include Hawk Cliff (Field, 1971; Haugh, 1972) and Point Pelee (Gray, 1961).

After passing these areas, hawks apparently tend to disperse widely although the Gulf Coast evidently acts as another important diversion-line, funneling Broad-winged Hawks around the Gulf of Mexico through Mexico into Central and South America. Ospreys and Peregrine Falcons, however, cross the Gulf and the Antilles. Some remain on the islands, but others continue into northern South America. In a few instances Ospreys penetrate well into the continent. One bird, reared and banded along coastal New Jersey, was recovered at Lima, Peru (Joseph Jacobs, personal communication).

OTHER MIGRATION ROUTES

Various other areas also serve as migration routes and are used by hawks during autumn. These seem to have less influence as diversion-lines because significant geographic features are less evident. Nonetheless, they often produce good numbers of birds at some of the spots.

One route is located between the Atlantic coast and the Appalachian Mountains. Some typical lookouts in this area are Hook Mountain and Mount Peter in New York, the Montclair Hawk Lookout Sanctuary and the Bearfort Fire Tower in New Jersey, and Pipersville, Pennsylvania. Most of these spots are on secondary ridges which extend for limited distances. Other lookouts are scattered throughout New England, north or west of the Kittatinny Ridge in Pennsylvania (e.g., the Pulpit on Tuscarora Mountain), and in various locations south of the Mason-Dixon line.

BROAD-FRONT MIGRATIONS

Some hawk watchers long have suspected that hawks migrate over broad fronts during autumn, that these birds are influenced by wind drift to a considerable degree, and that concentrations of hawks seen at the well-known lookouts are migration artifacts resulting from hawks coming into close contact with prominent geographic features which then act as diversion-lines. However, an adequate explanation of the phenomenon has been confused in various adjunct discussions. Broun (1949a), for example, discussed parts of the phenomenon, and Heintzelman and Armentano (1964) suggested a broad-front migration as a Trans-Mountain Drift hypothesis.* Their idea was partly correct, but it failed to recognize the important role which mountains, coastlines, or shorelines of other large bodies of water can have upon species such as Broad-winged Hawks, and apparently Sharp-shinned Hawks, by acting as diversion-lines.

DIVERSION-LINES

Murray (1964) was the first to develop a formal diversion-line concept for eastern North American hawk migrations, at least insofar as Sharp-shinned Hawks migrating along the Atlantic coastline are concerned. However, some of his ideas were questioned by Mueller and Berger (1967),† who contended that wind drift plays an important role in concentrating hawks at the well-known lookouts under certain wind-direction conditions. In my view, major geographic features act as diversion-lines for *limited distances,* and combine with wind drift to produce the concentrations of hawks at spots such as Bake Oven Knob and Hawk Mountain, Pennsylvania, and Cape May Point, New Jersey—particularly when northwesterly winds occur. There may be instances when this combination of factors fails to produce large hawk flights, but in general this appears to me to be the key to most large hawk flights using eastern North America's major hawk migration routes during autumn.

GEOGRAPHY, WIND DRIFT, AND DIVERSION-LINES

The relationships of geography, wind drift, and diversion-lines to broad-front autumn hawk migrations are complex. They are best

---

* A name applied to an explanation of the possible mode of operation of broad-front migrations.
† Murray (1964) and Mueller and Berger (1967) differ on the use of the term *diversion-line* versus *leading-line.*

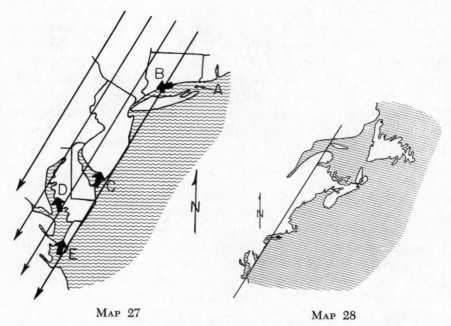

MAP 27   MAP 28

MAP 27   Diversion of migrating Sharp-shinned Hawks has occurred at various
coastal sites in the northeastern United States (short thick arrows): (A)
Fishers Island, N.Y.; (B) New Haven, Conn.; (C) Cape May Point, N.J.;
(D) Hooper Island, Md.; and (E) Cape Charles, Va. At Fishers Island,
N.Y., hawks continue southwestward toward Long Island. The general direction of the broad-front migration is indicated by the long arrows pointing
southwestward. Reprinted from Murray (1964).

MAP 28   The line extending northeast to southwest along the coast indicates
the limit of the eastern flank of the *bulk* of the migration of Sharp-shinned
Hawks according to the diversion-line theory. Reprinted from Murray (1964).

studied by considering each major migration route separately. Generally the most important factors which relate to hawk flights are
best studied locally or on a limited regional scale. At least five important factors combine with geography and wind drift to influence
broad-front hawk migrations along diversion-lines: (1) general
weather systems, (2) local weather conditions, (3) the altitude of
the flights as they approach prominent geographic features, (4)
the influence of thermals upon some species, and (5) the morphology and anatomy of the various species. The last two factors,
as previously discussed, are critical influences in precipitating

Broad-winged Hawk flights at some lookouts. Wind drift, of course, is a function of local and general weather conditions.

ATLANTIC COASTLINE

North of Connecticut, Atlantic coastal hawk migrations are poorly understood. Except for observations made at a few spots, notably Brier Island, Nova Scotia (Hawkes, 1958; Tufts, 1962; Wickerson Lent, unpublished data), very little systematic hawk counting has been done along the coastline of northern New England and Canada.

In addition to the Brier Island observations, a variety of hawks have been seen crossing Grand Manan Island (Squires, 1952). It would be interesting to know, therefore, if hawks also cross Machias Seal Island off Cutler, Maine. Southwest of Machias Seal, hawk

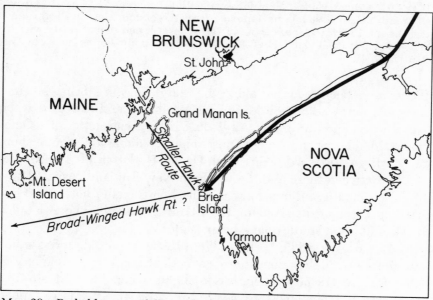

MAP 29   Probable routes followed by hawks migrating through Nova Scotia in autumn based upon information provided by Wickerson Lent. Smaller hawks and passerines fly northwest from Brier Island toward Grand Manan Island and Maine, whereas Broad-wings have been observed following a westward or west-southwestward course from Brier Island. These Broad-wing routes are subject to further study, however.

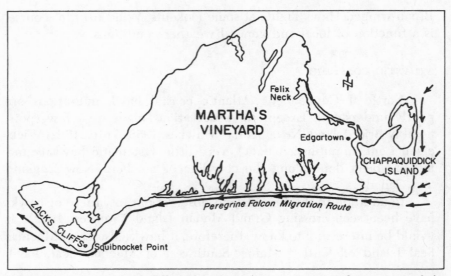

MAP 30 Autumn routes followed by migrating Peregrine Falcons at Martha's Vineyard, Mass. Most falcons approach from the eastern side of Chappaquiddick Island, follow the south shore of the Vineyard, and depart in the vicinity of Squibnocket Point and Zacks Cliffs. Based upon information provided by Gus Ben David II.

flights have been recorded along the Maine coastline from the summit of Mount Cadillac on Mount Desert Island and at sea between Nova Scotia and the Island (William C. Townsend, letter of 1 October 1973). Peregrine Falcons also migrate along the south shore of Martha's Vineyard (Gus Ben David II, letter of 12 October 1973). Between there and the southern New England coast, I know of no systematic autumn hawk counts being made, however. Connecticut is the next spot along the Atlantic coastline where detailed hawk migration studies have been made.

As early as 1885, Trowbridge (1895) reported and studied autumn hawk flights along the Connecticut coast in the vicinity of New Haven. He noticed increased flights when lowered air temperatures and northwest to north winds occurred. Sharp-shinned and Broad-winged Hawks were especially abundant, but other species also were seen. He writes: "All of the southern border on Connecticut is washed by Long Island Sound, and the entire shore lies nearly in an east and west direction. When the migrating hawks flew southward with the strong northerly winds, and arrived at the Sound, rather than fly over the water, they would turn westward and proceed along the coast until they arrived at the State of New York, where they would continue southward, through New Jersey and Pennsylvania." The hawks were subject to wind drift and, upon reaching the Connecticut coast, the majority flew westward using the coast as a diversion-line. He suggested, however, that many hawks may fly over the waters of Long Island Sound. The observations of Ferguson and Ferguson (1922) on Fishers Island confirm this idea, although the island is at the extreme eastern end of Long Island Sound and would not receive hawks crossing the Sound at New Haven. "Thus," as Trowbridge (1895: 269) stated, "it seems as if in the case of the flights in southern New England, that the east and west direction of the coast line, and the wind, both have their effect in influencing the migration of the hawks and other land birds."

The hawk flights at New Haven reflect birds drifting toward the coast from *northwestern* New England, whereas the Sharp-shin and other hawk flights recorded crossing east to west over Fishers Island at the eastern end of Long Island Sound (Ferguson and

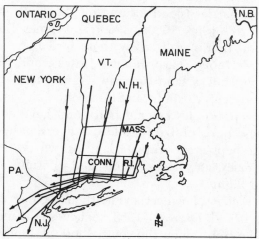

MAP 31   The general direction of coastal New England hawk migrations in autumn on prevailing northwest winds. Hawks concentrate at various coastal sites, particularly in the vicinity of New Haven, Conn. Redrawn from Trowbridge (1895).

Ferguson, 1922) demonstrate a neat connecting link with broad-front flights originating in *northeastern* New England and Canada. Thus, in southern New England and eastern New York, the major Atlantic coastline hawk migration route is split into two subroutes —especially in respect to accipitrine flights. Along both subroutes, shorelines serve as diversion-lines in connection with hawks involved in wind drift. The most important of the subroutes is the actual mainland (Trowbridge, 1895). The second subroute involves accipiters and falcons coming southward from northeastern New England. Fishers Island is the first point of land on this subroute's barrier-island diversion-line. The Sharp-shin flights occur on Fishers Island principally on northwest winds although some Marsh Hawks drift through daily during autumn regardless of wind direction. However, Peregrine Falcons and Pigeon Hawks ". . . prefer a south-west wind to any other. A strong wind is no hindrance to them, and we have come to feel that a typical Duck Hawk (Peregrine) day is one when the wind is blowing from the south-west, with almost a hint of bad weather" (Ferguson and Ferguson, 1922: 494).

Southwest of Fishers Island, hawks encounter long and relatively narrow Long Island. There the picture becomes more complex, but the field studies conducted by Darrow (1963) on Fire Island, a narrow barrier island located off the south shore of Long Island, and Ward (1958, 1960a, 1960b) at Jones Beach, help to trace the general routes utilized by these birds.

Accipiters, notably Sharp-shinned Hawks, and Marsh Hawks, Ospreys, Pigeon Hawks, and Sparrow Hawks tend to follow the Fire Island shoreline crossing the Fire Island inlet and continuing along Jones Beach. Studies beyond that point have not been made, but most of these birds probably follow a diversion-line route along Long Island, perhaps cross Staten Island as well, then enter northeastern New Jersey at various points and continue southward as fairly scattered flights.

Peregrine Falcons, however, evidently use a different route upon reaching Long Island. Darrow (1963) believes that most Peregrines follow the beach front along the south shore of Long Island for as much as one hundred miles, moving generally west. After leaving Democrat Point on Fire Island, however, they sometimes make a 30- to 45-degree change in direction, by-pass the New Jersey coast,

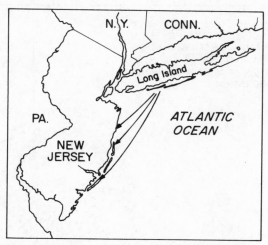

MAP 32 Probable autumn migration routes followed by shorebirds and some Peregrine Falcons after leaving Democrat Point, Fire Island, N.Y. Other Peregrines may alter their course at sea and either come ashore at Cape May Point or bypass New Jersey completely and continue on an over-the-water route until they reach the barrier islands along the Delmarva Peninsula, where concentrations of Peregrines occur in autumn. Redrawn from Darrow (1963).

and continue over the water to the Delmarva peninsula. He states that falconers trapping Peregrines arriving on the Delmarva beach front note that the birds appear to be tired and hungry. However, Darrow doubts if falcons arriving on Fire Island from the north have used substantial over-the-water flight. This explanation may be correct although Ward (1960a) reports excellent Peregrine, Pigeon Hawk, and Sparrow Hawk flights at Jones Beach on northwesterly winds but lesser flights on easterly winds. Elliott (1960) also traces some Long Island falcon flights along the Sound shore, and further suggests that the Pigeon Hawks and Sparrow Hawks also may leave the south shore in the vicinity of Rockaway Point to engage in a shorter over-the-ocean flight toward coastal New Jersey.

South of Long Island in New Jersey, Edwards (1939) mentioned some coastal falcon flights north of the Cape May peninsula. Major autumn hawk flights occur at Cape May Point at the extreme southern tip of New Jersey, however. .

In 1922, Witmer Stone wrote of the concentrations of hawks and songbirds at Cape May Point, noting that the largest flights

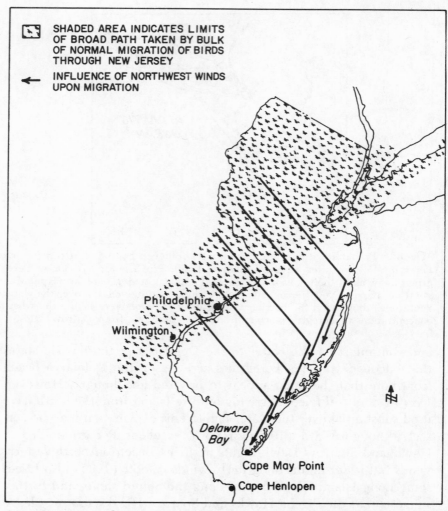

MAP 33 On northwest winds substantial numbers of hawks appear to be subject to wind drift (arrows) until the New Jersey coast is reached. The birds then follow the coastline southward as a diversion-line eventually concentrating at Cape May Point. The shaded area indicates the normal limits of most broad-front bird migrations through New Jersey. Redrawn from Allen and Peterson (1936).

occurred when northwest winds prevail. Allen and Peterson (1936) made a detailed study of hawk flights at the Point. They determined that hawks are subject to wind drift, and that concentrations of hawks seen at the Point ". . . seem to be very largely the result of a wind condition—a northwest wind blowing across the lane of travel. The birds lose ground against this wind and gradually slip into the southern New Jersey peninsula. These birds eventually jam into the narrow confines of Cape May Point. A north wind will bring birds, and even a north-northeast wind will bring a few, but a northwest wind almost invariably brings a great many more.

"The importance of the degree of wind force is illustrated by the observation that clear skies and light northwest winds bring only moderate flights while a northwest wind of fairly strong, or strong force is almost certain to be accompanied by a large influx of migrant Hawks besides many smaller land birds." Apparently the coastline serves as a diversion-line after hawks have drifted into close proximity to it. Murray's (1964) alternate explanation is discussed later in this chapter.

Upon reaching Cape May Point on northwesterly winds, hawks fly northward along the Delaware Bay side of the peninsula. Allen

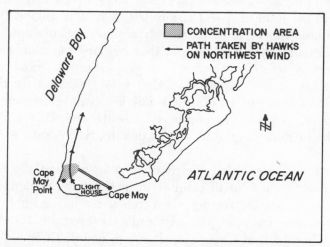

MAP 34 On northwest winds hawks concentrate at Cape May Point, N.J., then divert northward along the New Jersey side of Delaware Bay. Ultimately the birds presumably cross the bay at a relatively narrow, but still undetermined, spot. Redrawn from Allen and Peterson (1936).

and Peterson (1936: 397–399) followed hawks northward along the Bay for seventeen miles and state that Richard H. Pough observed considerable numbers crossing near Wilmington. When winds shift to another quarter, however, hawks cross Delaware Bay from the tip of the Point. Sharp-shins usually fly very high when crossing. However, repeated efforts to detect these flights at Cape Henlopen, Delaware, toward which they seem to be headed, have failed. Allen and Peterson (1936: 399) suggest that the hawks still are very high upon reaching Delaware ". . . and spread out over the country-side before dropping down." Even today, no substantial evidence exists to demonstrate the fate of the hawks after they leave Cape May Point. Nonetheless, some insights into the mechanics of these flights are known.

In 1937, Witmer Stone made significant comments about the altitudes of the flights crossing the Cape. "As a result of many years of observation I have come to the opinion that the normal southward flight is at a great height, just as I saw it fifty years ago in the vicinity of Philadelphia . . . and that it is only when the northwest gales threaten to drive the birds out to sea that they descend and head into the wind to save themselves and, incidentally, cause the well known visible flights; at other times they are passing regularly, but beyond the limit of man's vision. If this is so it follows that the hawks counted during a flight are only a relatively small proportion of those actually on migration and that perhaps the number killed by gunners is relatively smaller, in proportion to those passing through, than we had supposed, which would account for the fact that several species of hawks have not yet been exterminated although greatly reduced in numbers!" Choate and Tilly (1973: 1) confirm the extreme heights at which some hawks migrate over Cape May Point.

Stone's suggestion that the number of hawks actually counted may represent only a small proportion of the number actually migrating is extremely interesting when considered in the light of some recent data. For example, the extraordinary Sparrow Hawk flight of 16 October 1970 at Cape May Point (Choate, 1972) adds considerable weight to Stone's statement. Moreover, the spectacular Goshawk invasion in the eastern United States during the autumn of 1972 further suggests that current concepts of raptor populations may have to be revised upward considerably. The next chapter details more information on raptor populations and the possible relationship of autumn hawk counts to them.

South of New Jersey, understanding of Atlantic coastal hawk flights is incomplete and somewhat confusing. Data from Delaware (Mohr, 1969) still are very fragmentary. Better known are the Peregrine Falcon migrations at Assateague Island off coastal Maryland and Virginia (Berry, 1971; Ward and Berry, 1972; J. L. Ruos, unpublished data). Not until one reaches the Cape Charles peninsula along coastal Virginia are detailed hawk migration studies again available. These flights were studied during the autumn of 1936, just south of the town of Kiptopeke, by William J. Rusling (1936). This spot is at the extreme southern end of a long, narrow peninsula similar to the Cape May peninsula.

Rusling (1936) recognized that wind and geography combine to cause concentrations of hawks and other birds. With northwesterly winds gradually drifting hawks toward the Atlantic coastline, the birds eventually follow the coastline southward. At the neck of the Cape Charles peninsula, at the southern boundary of Maryland, there is a narrowing of the peninsula to a width of about ten miles. High-flying birds, seeing Chesapeake Bay to the west, apparently prefer to fly into the wind when near or over large bodies of water. They turn to the westward, then northwestward, into the wind blowing down the Bay coastline. At Hooper Island, concentrations of hawks and other birds occur.

Rusling (1936: 11) reports that northeast winds always bring large numbers of migrants to Cape Charles. These winds aid hawks crossing the coastal area by bringing them into the peninsula's neck, then southward into the peninsula, ultimately to produce the great concentrations just south of Kiptopeke. That is, northwest winds produce hawk flights at the northeastern part of the Cape Charles peninsula whereas the southern end gets them on northeasterly winds.

Once hawks are committed to migration down the peninsula, they fly toward the vicinity of the Point regardless of wind direction. At the tip of the peninsula, however, they do not cross the mouth of Chesapeake Bay on a tail wind unless it is exceedingly light. Instead, they drop to a lower altitude and fly northward into

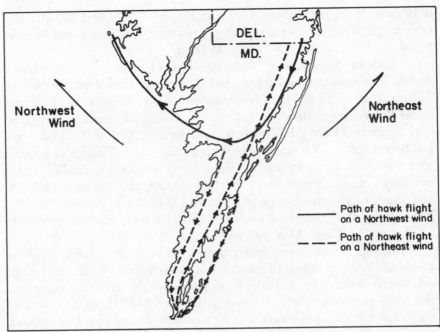

MAP 35 Routes followed by hawks approaching the Cape Charles, Va.,
peninsula on northwest and northeast winds during autumn. Redrawn from
Rusling (1936).

the wind. But with a full or partial head wind, the hawks unhesitat-
ingly cross the bay.

Rusling (1936: 21–33) discovered that hawks use two distinct
flight lines as they migrate over the southern part of the Cape
Charles peninsula: (1) southward over the mainland, and (2)
southward over a chain of outer islands paralleling the coast. The
islands curve in a southwestward direction near the end of the peninsula with Fisherman's Island, the last in the chain, positioned a
mile due south of the peninsula tip. Rusling considered Fisherman's
Island as the focal point of the entire Eastern Shore hawk flights
before they continue across the mouth of Chesapeake Bay. The
maps illustrate the paths followed by migrating hawks upon reaching the tip of the peninsula under various wind directions. Not illustrated are the paths used over the outer islands upon reaching
Chesapeake Bay on prevailing northerly winds. Birds reaching

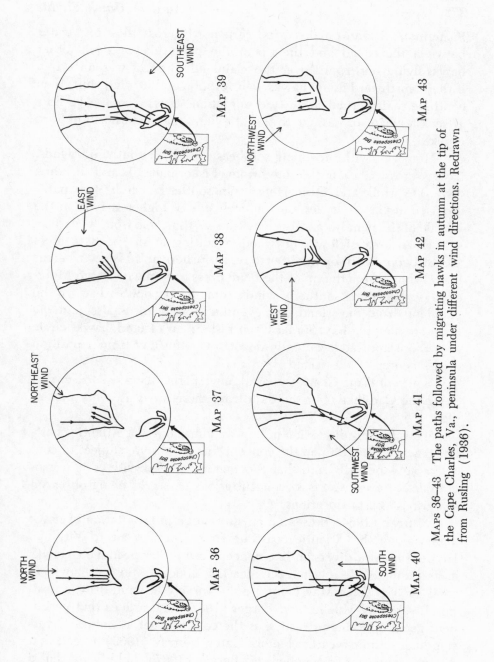

MAPS 36–43   The paths followed by migrating hawks in autumn at the tip of the Cape Charles, Va., peninsula under different wind directions. Redrawn from Rusling (1936).

Fisherman's Island on this wind turn north, cross the open water between the island and the tip of the peninsula, and join other hawks flying northward over the mainland. Rusling was uncertain if these northward flying hawks wait for a favorable wind shift, then continue their normal southward migration across Chesapeake Bay, or seek a narrower passage to the Maryland mainland north of Kiptopeke.

However, with prevailing southeast, south, or southwest winds, hawks cross Chesapeake Bay from Fisherman's Island heading toward Norfolk and Cape Henry. On southerly winds, two paths are used at Cape Charles, one on each side of higher ground in the middle of the peninsula. Most hawks leave the Cape from its eastern side at an area of low pine woods resulting in the need to cross almost two miles of open water to reach Fisherman's Island, whereas birds leaving from the tip of the Point cross one mile of water. Moreover, on prevailing southerly winds, some hawks double back to the mainland from the island, but eventually these birds also join the flight crossing the Bay. Similarly, on Fisherman's Island, hawks circle over the island but eventually depart over the Bay from a position in the center of the island.

South of Cape Charles, coastal hawk flights have been recorded along the Carolinas, but the data from these areas are very incomplete.

Thus far this discussion of hawk flights along the Atlantic coastline has been based upon the concept that hawks are subject to considerable wind drift, and that this factor, coupled with certain geographic features, results in concentrations of hawks being observed at various coastal locations.

Murray (1964) presented an alternative interpretation of hawk flights along the Atlantic coastline from Connecticut to Virginia. He developed a diversion-line concept using Sharp-shinned Hawks as an example. He rejected the idea that concentrations of hawks at spots such as Cape May Point are due largely to wind drift, a position favored by Mueller and Berger (1967), and argues that prominent geographic features such as the coastline act as natural diversion-lines despite weather factors. Later, Murray (1969) agreed that his idea, and that of Mueller and Berger (1967), hold ". . . that a northwest wind is most effective in causing concentrations; the one, by drifting the hawks, the other, by diverting hawks."

In point of fact, a diversion-line usually exerts an important influence upon migrating hawks when the birds are subject to wind drift. That is, wind drift and diversion-lines combine to produce the notable autumn hawk concentrations at the well-known eastern North American spots. This is well illustrated inland along ridges such as the Kittatinny. However, on the ridges, as well as along the coast, large flights of hawks occur regularly only along certain sections of these geographic features.

INLAND MOUNTAINS

Inland from the Atlantic coastline, numerous mountain ranges serve, in varying degrees, as diversion-lines for migrating hawks. Most of these ridges are of secondary importance as flyways. However, a few serve as important diversion-lines and hence are major autumn migration routes. It is perhaps significant that the most important mountain flyways are oriented in a general northeast to southwest axis and are relatively unbroken for long distances.

In New England, hawk migrations still are too incompletely studied to permit adequate evaluation of most mountain ranges as diversion-lines, but available data suggest that most ridges probably can serve as limited diversion-lines and flyways. A notable exception to this fragmentary status of New England hawk migration studies is the work done at Mount Tom in the Tom Range near Springfield, Massachusetts. Hagar (1937a, 1937b) established the importance of this range as a major hawk migration flyway soon after Hawk Mountain, Pennsylvania, was discovered, and Goat Peak in Mount Tom State Reservation still is New England's best hawk lookout. Elsewhere, as in southern New England, the studies of Hopkins and Mersereau (1971, 1972) ultimately may reveal other ridges of equal importance.

South of New England, there are some particularly notable examples of geographic features serving as diversion-lines in combination with wind drift and other factors. Foremost are sections of the Kittatinny Ridge in northwestern New Jersey and eastern Pennsylvania. When northwesterly or northerly winds occur, hawks are diverted along this ridge for varying distances before again becoming subject to wind drift and broad-front migrations. As hawks follow the ridge they pass several lookouts strung like beads along 125 miles or more of the ridge crest. This provides an unparalleled opportunity to study the effect of a major diversion-line upon autumn hawk flights.

Raccoon Ridge, New Jersey, is the most northeasterly of the major Kittatinny Ridge lookouts (Edwards, 1939; Tilly, 1972a, 1972b, 1973). All hawk species normally seen elsewhere along this mountain occur at Raccoon Ridge. Red-tailed Hawk flights are particularly impressive. Northwest or north winds are most favorable for Red-tail flights. For example, each of the four Red-tail flights exceeding 400 birds in November 1971 occurred when northwest or north-northwest winds prevailed (Tilly, 1972b: 3). Similarly, in November 1972, six Red-tail flights exceeded 100 birds. Of these 1 occurred on a north wind, and 5 on northwest or north-northwest winds (Tilly, 1973: 7). Although sizeable flights of other species sometimes occur when winds from other directions prevail at Raccoon Ridge, strong winds with northerly components usually are most desirable at this spot.

Raccoon Ridge is unusual among Kittatinny Ridge lookouts in that the Delaware River flows parallel and adjacent to its north slope. Moreover, Tilly (1972a: 22) points out that a parallel secondary ridge runs along the north side of the river. This lower ridge ". . . probably diverts a goodly number of hawks away from Raccoon." Hence the main fold of the Kittatinny, the parallel secondary ridge, and the Delaware River may serve as diversion-lines for migrating hawks. Perhaps the three enhance each other, especially for Bald Eagles and Ospreys, when north or northwest winds occur. However, the main fold of the Kittatinny almost certainly is the most important of the three geographic features insofar as their diversion-line roles are concerned. The sizeable hawk flights which occur at other lookouts southwestward along the ridge, and which lack any, or major, riverine features support this view. So, too, does the lack of evidence of sizeable hawk flights following the Delaware River southward for extended distances. However, during September of 1972, while evaluating the relative importance of Tott's Gap a few miles southwest of Raccoon Ridge (Heintzelman, 1973), I

spent a morning at a lookout on the west side of Delaware Water
Gap and watched an Osprey appear over the Delaware River north
of the Gap. Instead of following the Kittatinny Ridge southwestward, the bird followed the river downstream, through the Gap,
and continued out of sight. I was unable to determine if the hawk
was using a limited section of the river as a feeding ground prior
to continuing its migration along the ridge, or if the bird followed
the river for an extended distance.

At times the river probably serves as a potentially important
food source for migrating Bald Eagles and Ospreys as they pass
Raccoon Ridge. Indeed, hawk watchers sometimes observe Ospreys
flying past this spot, and other ridge lookouts, carrying fish in their
talons. This suggests that the hawks sometimes temporarily halt
their migrations briefly to hunt over rivers such as the Delaware, or
other aquatic habitats. But the extent to which the Delaware and
other rivers serve as important food sources and diversion-lines for
migrants is largely unknown except for fragmentary and inconclusive data.

Bake Oven Knob, north of Allentown, Pennsylvania, is the next
major lookout along the Kittatinny Ridge. There is no water course
in close proximity to this site. However, either of two lookouts are
available and used depending upon prevailing wind direction. When
northerly or westerly winds occur, most hawks fly along the north
side of the ridge, and observations are made from the North Lookout. The South Lookout is used when hawks fly along the south
slope under prevailing easterly and southerly winds (Heintzelma
and Armentano, 1964; Heintzelman, 1972d). About one and o
half miles southwest of Bake Oven Knob, Bear Rocks is another
standing lookout (Heintzelman, 1972d). Hence both spots are
able monitoring points permitting comparisons of hawk flight
this section of the Kittatinny Ridge.

Southwest of Bake Oven Knob and Bear Rocks, Haw
tain Sanctuary is the most famous hawk lookout in Americ
1949a). Its North Lookout is exceptionally productive v
west or northerly winds occur, but lesser flights usually
other winds. However, the establishment of a new
distance south of the North Lookout demonstrated t
of the hawks and eagles" sometimes were being o

North Lookout (Nagy, 1967: 5). Complicating the role of the Kittatinny Ridge as a diversion-line at Hawk Mountain, however, is the presence of the Little Schuylkill River flowing perpendicular to, then parallel to, the ridge. This is a minor tributary, but it may occasionally lead Ospreys, for example, to the ridge from areas north of the Kittatinny.

West of the Susquehanna River, the next important Kittatinny Ridge lookout is Sterrett's Gap. This site is no longer used as an observation post, but the earlier work of Frey (1940, 1943) demonstrated that it was a productive spot. A short distance southwest of Sterrett's Gap, hawk counts have been made for many years from Waggoner's Gap (Carlson, 1966; D. L. Knohr, personal communications). As discussed earlier, hawk flights passing these two sites often contain substantial numbers of birds not seen at Hawk Mountain or at other Kittatinny lookouts northeast of Hawk Mountain.

Early studies at Hawk Mountain (Broun, 1935, 1939; Poole, 1934) suggested that the Kittatinny Ridge serves as an extended flyway allowing hawks to utilize updrafts and follow the ridge for hundreds of miles. Broun (1935: 235), for example, stated: "These currents of air enable the birds to coast for mile on mile; thus they enjoy easy transit to their winter feeding grounds." Exceptions occur, however. Thus Broun (1939: 431) mentioned an example of groups of Broad-wings coming ". . . out of the north (at very great height), as though from adjacent ridges, and then proceeded to follow our ridge." He also mentions that hawks tend to avoid Hawk Mountain on easterly and southerly winds.

Nonetheless, early studies at Hawk Mountain produced major new hawk migration information. However, these data were collected at a single station. Hence they could not reveal, except indirectly, the extent or limitations of the importance of the ridge as diversion-line for migrating hawks. It was clear to several observers that comparative field studies, made from several locations along Kittatinny Ridge, would be needed to understand adequately movements of migrant hawks along the ridge. Thus, a few years Hawk Mountain was established, several other lookouts were various observers in an attempt to gather additional information on the passage of migrating hawks in northwestern New and eastern Pennsylvania.

Raccoon Ridge, New Jersey, was one of the first sites to be used as a hawk lookout late in the 1930's (Edwards, 1939). Unfortunately few data were published although some unpublished information gathered by William Rusling recently has come to light. Thus Edward Snively Frey's (1940, 1943) careful hawk counts in 1938 and 1939 at Sterrett's Gap, Pennsylvania, appear to be the first published *systematic* attempts to trace hawk flights along the Kittatinny Ridge after the birds passed Hawk Mountain.

Frey (1940) discovered great differences when Sterrett's Gap records were compared with Hawk Mountain data due to the presence of individual birds *not* seen at Hawk Mountain, and the absence of large numbers of hawks (perhaps 70 percent of the Hawk Mountain flight) drifting from the ridge between the two points. He pointed out that the number of hawks counted at Hawk Mountain is 35 percent greater than at Sterrett's Gap. Broun (1961: 8–9), however, later demonstrated that this is not always true by comparing hawk counts from Waggoner's Gap (just southwest of Sterrett's Gap) and Hawk Mountain. Nonetheless, Frey (1940) noted that the differences in the two station's counts were caused by ". . . the greater numbers of the most-abundant species at Hawk Mountain." Less abundant species gave a different picture. This was much more true for 1938 than 1939.

"The differences . . . point to birds not seen at Hawk Mountain only when the Sterrett's Gap figure is very nearly equal to or greater than the Hawk Mountain figure. It is well known that many birds leave the ridge in the seventy miles between the two points, a factor which would account for the disparity in numbers at the second point. But where the figures very nearly equal or exceed those of Hawk Mountain the presence of other birds would certainly be indicated.

"This would not be convincing evidence were it not for another discovery, namely, in the study of two species, Golden Eagles and Duck Hawk, it is evident that possibly 70% or better of the individuals of these species seen at Sterrett's Gap are not seen at Hawk Mountain. I am fully aware that this is an astonishing statement but the observations that follow leave me no choice but to make it. The typical case is that of the Golden Eagles for October 1, 1938. On that day the first Golden Eagles of the season for each place were tallied, four at Sterrett's Gap and four at Hawk Mountain. The eagles recorded at Hawk Mountain were: 1 adult at 9.15; 1 adult and 1 second-year bird at 4.10; 1 adult at 4.42. Those recorded at Sterrett's Gap were: 1 immature at 9.00; 1 adult at 9.52; 1 adult at 2.59; 1 second-year bird at 3.45. Three of these eagles could not possibly have been seen at Hawk Mountain. One, the adult tallied at Sterrett's Gap at 2.59, could have been the adult noted at Hawk Mountain at 9.15. Another similar case in point is the comparative analysis of the Duck Hawks recorded for the first eleven days of October 1939. At Hawk Mountain the total for those days was eleven, at Sterrett's Gap seventeen or at least six Peregrines at our point that were not observed at Hawk Mountain. The times and days for these birds compared present a hodge-podge as does every effort for the least-frequent migrants. There are numerous incidents such as these two pointing always to the belief that only a small percentage of the Sterrett's Gap birds have flown past Hawk Mountain.

"The best evidence of a differently constituted flight at Sterrett's Gap is found from a comparison of flight graphs depicting the intensity of flight at both places correlated with the dates. If the majority of Sterrett's Gap birds were the same as those of Hawk Mountain, the great flights there should affect the intensity of the flight at Sterrett's Gap on the same or the following day. Such seems to be generally the case but there are numerous and important exceptions. If a great flight should occur at Sterrett's Gap before one occurs at Hawk Mountain, then obviously the flight had little or no connection with Hawk Mountain since the migrants must pass Hawk Mountain before reaching Sterrett's Gap. On October 5, 1939, 1480 raptors flew over Sterrett's Gap. For the same day Hawk Mountain produced 439 birds with a total for the three preceeding days of 440 as against the Sterrett's Gap total for the same three days of 680. Assuming that all of Hawk Mountain birds were seen at our point (which is an absurd assumption on the grounds of every record we have) there are left 1281 hawks that were never tallied at Hawk Mountain. There are six other incidents of this same thing for the two Octobers of 1938 and 1939 alone."

Frey concluded that ridges north of the Kittatinny are the

sources of hawks observed at Sterrett's Gap but not at Hawk Mountain. However, more birds drift from the ridge between Hawk Mountain and the Gap than are added to the Gap's counts. The Sterrett's Gap figures also demonstrate that the northern ridges also act, to some extent, as diversion-lines for migrating hawks. Recent hawk migration studies at Tuscarora Mountain, Pennsylvania, also demonstrate this (C. L. Garner, personal communication).

After the termination of the Sterrett's Gap studies, little effort was made to continue to compare Kittatinny Ridge hawk flights. In *Hawks Aloft,* for example, Broun (1949a) emphasized a concept of a ridge migration without mentioning Frey's conflicting data although specific attention was called to the tendency for hawks to leave the ridge and scatter when calms or light winds occur other than from the north or northwest.

George W. Breck (1959, 1960a) apparently was the next person to attempt to trace routes used by migrating hawks in the New York-New Jersey-Pennsylvania region. His efforts were short-term projects using observers at many lookouts in several states for two or three days each autumn. The results were largely inconclusive due to data limitations. Refer to the original reports for full details.

In 1961, I began systematic field studies at Bake Oven Knob, Pennsylvania, in another effort to make comparative hawk counts along a portion of the Kittatinny Ridge. The initial results, although based upon a limited number of observation days, were similar to the conclusions reached by Frey (1940) at Sterrett's Gap. That is, it was shown that great variations sometimes exist between hawk counts made at Bake Oven Knob and those made at Hawk Mountain (Heintzelman and Armentano, 1964). For example, on 14 September 1963, 837 Broad-wings were counted at the Knob prior to eleven o'clock whereas Hawk Mountain reported 3,464 by ten o'clock and a total of 3,949 by noon. However, another 2,800 Broad-wings passed Bake Oven Knob between 1100 and 1710, whereas no hawks were seen at Hawk Mountain between noon and 1315 and only 1,307 were counted during the afternoon. But at the Pinnacle, four miles south of Hawk Mountain, an additional 1,540 Broad-wings were seen in the afternoon. These birds, *not counted at Hawk Mountain,* nonetheless were added to the Hawk Mountain count for

14 September. The probable mechanics of this flight were explained by Heintzelman and Armentano (1964) via a Trans-Mountain Drift hypothesis.

In recent years, additional field studies of hawk migrations at Bake Oven Knob, when compared with similar studies made at Hawk Mountain, tend to confirm my earlier conclusions that counts from the two locations frequently exhibit marked differences in the numbers of birds seen at each location, and Frey's (1940) conclusion for Sterrett's Gap that different hawks are involved, at least in part, in many Kittatinny Ridge hawk flights. An examination of some examples of field data, for several species, supports these conclusions.

Since Broad-winged Hawks seem to exhibit the most marked discrepancies in daily counts made at Bake Oven Knob and at Hawk Mountain, some Broad-wing flights are examined first.

EXAMPLE NO. 1.—On 17 September 1966, observers at Bake Oven Knob counted 1,392 Broad-winged Hawks whereas 2,441 were counted at Hawk Mountain. Hence the flight at the Knob was about 57 percent of the magnitude of the Hawk Mountain flight. A notable feature of much of the Bake Oven Knob flight was that large kettles of Broad-wings were seen over the valleys north and south of the ridge suggesting a broad-front migration was underway. Few birds seemed to be using the Kittatinny Ridge as a diversion-line.

EXAMPLE NO. 2.—Light easterly winds on 12 September 1967 produced flights of 1,844 Broad-winged Hawks at the Knob and 2,530 at Hawk Mountain's North Lookout. In this instance the Bake Oven Knob flight was about 73 percent of the magnitude of the Hawk Mountain flight. My field notes state that a broad-front migration apparently was in progress. The hawks seemed to be drifting cross-country, using some sections of the Kittatinny Ridge as a diversion-line.

EXAMPLE NO. 3.—On 13 September 1967, light northerly winds in the morning shifted to calm to light southeasterly afternoon winds. At Bake Oven Knob we counted 2,575 Broad-wings whereas at Hawk Mountain's North Lookout 641 were seen, the latter representing about 25 percent of the Bake Oven Knob flight. Only 197 (about 7.5 percent) of the total Bake Oven Knob count occurred during the morning on north winds in contrast to 404 Broad-wings (63 percent of the flight) passing Hawk Mountain in the morning.

Afternoon winds shifted to the southeast at the Knob producing 2,378 Broad-wings (about 92.5 percent of the day's count). My field notes indicate that a massive Broad-wing flight developed over both valleys, north and south of the Kittatinny Ridge, after the wind shift. The hawks, using thermals, were drifting cross-country in a broad-front movement with relatively few birds using the ridge as a diversion-line. Hence large numbers were seen at one location upridge without ever approaching another downridge lookout.

EXAMPLE NO. 4.—Calm to very light and variable winds, coupled with considerable haze, occurred at Bake Oven Knob on 16 September 1968 where 2,489 Broad-winged Hawks were counted in contrast to only 576 being seen at Hawk Mountain's North Lookout. The latter count thus represents about 23 percent of the magnitude of the Bake Oven Knob flight.

Prior to 1500 hours, only 496 Broad-wings (about 20 percent of the day's total) passed the Knob whereas at the North Lookout at Hawk Mountain 365 Broad-wings (about 63 percent of the day's total there) were seen. At 1510 hours, however, a massive Broad-wing flight developed at Bake Oven Knob. At least six huge flocks (some containing over 500 birds) appeared, circled skyward in thermals, then glided southwestward along the north slope of the ridge in an apparent use of existing weak updrafts. This flight continued for nearly an hour resulting in 1,892 hawks being counted during this single one hour period. This is about 76 percent of the entire day's Broad-wing count.

Apparently due to nearly calm winds many birds drifted cross-country bringing large numbers directly over the Knob. Thousands of additional Broad-wings probably passed through the area undetected because of the extreme difficulty of seeing them against blue skies and haze.

EXAMPLE NO. 5.—Flight conditions at Bake Oven Knob on 17 September 1968 appeared even less favorable than on the previous day. Extreme haze restricted visibility to three miles, and calm winds prevailed for many hours. Nonetheless, the day's Broad-wing count reached 4,137 birds—the highest ever tallied at the Knob. Downridge, at Hawk Mountain's North Lookout, 4,863 Broad-wings were seen. The Bake Oven Knob flight therefore was about 85 percent of the magnitude of the Hawk Mountain flight. The composition of the flights at the two locations had some markedly conflicting patterns, however. The morning flight at the Knob contained only 119 Broad-wings or about 3 percent of that location's count. In contrast, 2,620 Broad-wings were seen during the morning at the North Lookout at Hawk Mountain—about 54 percent of the day's total. Although the afternoon flight at Hawk Mountain was large, containing 2,243 birds or about 46 percent of the day's total, the afternoon flight at Bake Oven Knob was spectacular. We tallied 4,018 birds or about 97 percent of the day's total. The bulk of this flight occurred during an interval of about 75 minutes. At 1445 hours a massive flight of hawks appeared unexpectedly. During a 15 minute period, 1,340 Broad-wings passed, and the next hour (1500–1600, EST) produced an additional 2,416 birds. At times the sky literally was covered with a wide stream of hawks gliding overhead.

The data for 17 September, although containing similar numbers of birds for both locations, clearly show that the flights seen at Bake Oven Knob were composed, at least in large measure, of different birds from those seen at the North Lookout at Hawk Mountain.

These are typical examples of large flights of Broad-winged Hawks observed at various Kittatinny Ridge lookouts. They demonstrate that many such flights are broad-front migrations, that different populations of birds frequently are seen at lookouts located in relatively close proximity to each other, and that the ridge frequently fails to act as a diversion-line for migrant Broad-wings—particularly when very light or calm winds occur. At other times, e.g., when moderately brisk winds occur, Broad-wings are diverted along the ridge for varying distances, and the counts at lookouts such as Bake Oven Knob and Hawk Mountain may be similar.

The influence of the Kittatinny Ridge as a diversion-line upon nonthermal-dependent hawks almost certainly may be different than it is for Broad-wings. Hence one should examine other species such as Sharp-shinned and Red-tailed Hawks since they, too, are common autumn migrants along the inland ridges. Despite the fact that Sharp-shins, for example, also are common migrants at certain Atlantic coastline concentration points, those birds forming the inland ridge flights are markedly dependent upon updrafts to aid them in completing their migrations. To determine if Sharp-shins make extensive use of the natural diversion-line features of the Kittatinny Ridge, thus resulting in similar counts at Bake Oven Knob and Hawk Mountain, some Sharp-shinned Hawk flights also are examined.

EXAMPLE NO. 1.—On 2 October 1966, calm to moderately brisk west winds prevailed at Bake Oven Knob where 115 Sharp-shinned Hawks were counted. In contrast, downridge at Hawk Mountain's North Lookout 211 Sharp-shins were seen. Hence the Bake Oven Knob count was about 55 percent of the magnitude of the Hawk Mountain count. Since nearly all of the hawks flew along the north slope of the ridge, apparently using it as a diversion-line, as would be expected on westerly winds, it seems likely, in my opinion, that additional birds joined the Kittatinny Ridge flight after drifting southward on westerly winds from the various parallel ridges north of Hawk Mountain. These mountains, which tend to bend toward Hawk Mountain and converge or terminate just north of Hawk Mountain, provide flyways for some migrating hawks although the flights apparently are modest in number. Nonetheless, the addition of hawks to the Kittatinny Ridge flight, some distance upridge from Hawk Mountain, might reasonably account for the higher count of Sharp-shins at the downridge location.

EXAMPLE NO. 2.—The Sharp-shin flight of 8 October 1966 apparently witnessed few hawks joining or leaving the Kittatinny Ridge as 190 hawks were counted at Bake Oven Knob and 221 were seen at the North Lookout at Hawk Mountain. Curiously, calm to light westerly winds prevailed throughout the day which, one would expect, might result in considerable numbers of birds drifting away from the ridge. Nonetheless, according to my field notes, most of the birds flew near eye level on both sides of the ridge. During periods of calms many hawks were seen along the south slope, but when a wind sprang up the birds flew along the north side again. Hence, at least for this flight, natural geographic features seemed to provide adequate diversion-line properties regardless of wind, or a lack of it.

EXAMPLE NO. 3.—On 16 October 1967 light southwest winds prevailed at Bake Oven Knob, and the count from the South Lookout was 178 Sharp-shinned Hawks. But downridge, at the South Lookout at Hawk Mountain, 333 Sharp-shins were reported—an astonishing discrepancy. Nonetheless, the evidence shows that the Bake Oven Knob count was only about 53 percent of the magnitude of the Hawk Mountain count.

Normally hawks fly along the south slope of the ridge with prevailing southerly winds and, not infrequently under such conditions, considerable numbers drift away from the ridge between Bake Oven Knob and Hawk Mountain. Apparently this flight was different, however. Few birds seem to have drifted from the ridge thus avoiding detection at Hawk Mountain. In point of fact, large numbers of new birds seem to have supplemented the flight somewhere between the two locations. Thus the evidence suggests that geographic factors, perhaps as much as wind, may influence Sharp-shinned Hawks to follow a prominent geographic feature and use it as a diversion-line.

EXAMPLE NO. 4.—The flight of 21 October 1967 produced more Sharp-shinned Hawks at Bake Oven Knob than at Hawk Mountain's North Lookout, however. At the former location 249 Sharp-shins were counted on light to moderate southwest to west winds, whereas 204 were counted at the latter lookout. Since the hourly distributional patterns of the flights are similar, one must conclude that few birds drifted from the ridge; the bulk of the flight followed the Kittatinny Ridge as a diversion-line between the two locations. Most of the birds passed at eye level, or a little higher or lower, and flew along the north slope of the mountain.

EXAMPLE NO. 5.—At the South Lookout at Bake Oven Knob on 2 October 1968, Robert MacClay recorded light south to southwest winds during the morning and brisk southwest afternoon winds. The Sharp-shin count was 179—more than double the 80 seen at Hawk Mountain's North Lookout. The evidence thus shows that about 55 percent of the Bake Oven Knob flight apparently either failed to follow the Kittatinny Ridge southwestward to Hawk Mountain or the hawks were too far south of the latter location's North Lookout to be seen. Regardless, one again is faced with another discrepancy —i.e., geographic features, by themselves, did not induce many of these hawks to follow a ridge diversion-line. Hence this example conflicts with Example No. 3.

EXAMPLE NO. 6.—The Sharp-shinned Hawk flight of 5 October 1968 occurred on moderate to brisk northwest winds. At Bake Oven Knob 207 were counted—only about 56 percent of the 371 Sharp-shins counted at Hawk Mountain's North Lookout. Several inter-acting factors may have combined to cause the discrepancy in the counts. First, some new birds from ridges north of the Kittatinny doubtless joined our ridge flight southwest of the Knob. In addition, the hawks rose higher and higher around noon which produced somewhat of a noon lull at the Knob but *not* at Hawk Mountain. Hence some hawks, passing the Knob beyond man's vision, may have dropped to a lower elevation and thus were counted at Hawk Mountain. It is unlikely that many Sharp-shins drifted away from the Kittatinny Ridge between the two lookouts, however, as north-west winds usually produce good ridge flights for at least limited distances.

The previous examples, typical of some of the variations in Sharp-shinned Hawk flights along the Kittatinny Ridge in eastern Pennsylvania, suggest that these small accipiters are subject to wind drift to some degree but seldom to the extent that characterizes many Broad-winged Hawk flights. Nonetheless, comparative data from Bake Oven Knob and Hawk Mountain show that some Sharp-shins join the ridge flights and others leave them. The extent to which this occurs varies considerably from day to day, largely de-pending upon wind conditions. At times the numbers of birds join-ing or leaving a flight can be considerable. Despite these exceptions the bulk of the Sharp-shinned Hawks passing Bake Oven Knob tend to follow the ridge at least as far as Hawk Mountain before gradu-ally drifting southward from the ridge to be replaced by new birds which themselves have drifted toward the Kittatinny Ridge and joined flights using that geographic feature as a diversion-line.

In 1971, increased interest in hawk migrations developed in the northeast (Anonymous, 1972). Thus it was possible to complete intensive autumn-long field studies at three or more locations along the Kittatinny Ridge in Pennsylvania and New Jersey. Data were gathered by Sharadin (1972) at Hawk Mountain, Pennsylvania, by Heintzelman and MacClay (1972) at Bake Oven Knob, Pennsylvania, and by Tilly (1972b) at Raccoon Ridge, New Jersey. These studies permitted a comparison of daily hawk counts from three major lookouts spaced about fifty-five miles along the ridge. Final conclusions cannot be made until more data are available but, in making a preliminary study of selected Sharp-shin flights, Heintzelman (1972f) suggested that a trend toward an increase in the number of Sharp-shinned Hawks sometimes occurred as one progressed from lookout to lookout southwestward along the Kittatinny Ridge between Raccoon Ridge and Hawk Mountain. Exceptions occurred on several days, however. In 1972, data gathered by Sharadin (1973), by Heintzelman and MacClay (1973), and by Tilly (1973) for the same three lookouts again suggested, despite some marked contradictions, that when large Sharp-shinned Hawk flights occur, the ridge is used as a diversion-line resulting in an increasing number of birds being seen at the respective lookouts southwestward along the Kittatinny Ridge (table 56). However, exceptions again occurred, and more data are needed to verify such a general trend on a yearly basis. Nonetheless, the preliminary data suggest that more hawks drift to the ridge and join the flight than leave, thus accounting for the apparent increase in numbers of Sharp-shins often counted at major lookouts southwestward along the ridge. Beyond Hawk Mountain, however, more hawks apparently leave the ridge than join it (Frey, 1940).

Although the previous information is fascinating, and ultimately may be verified as being essentially correct for most hawk flights, it must again be emphasized that many years of additional field study are necessary before the subtle details of Sharp-shinned Hawk flights can be understood fully. Unfortunately, not all field studies produce data directly related to the main problem. Thus the hawk counts made at the Catfish Fire Tower, New Jersey (Heintzelman, 1972a), and at Tott's Gap, Pennsylvania (Heintzelman, 1973), although of value and interest in themselves, failed to produce good data of direct use in studying the main Kittatinny Ridge hawk flights since neither site was of major importance due to location and/or topography.

The Red-tailed Hawk is the third autumn migrant which occurs in large numbers along the Kittatinny Ridge. Hence an examination of some Red-tail flights is worthwhile to evaluate the Kittatinny Ridge's diversion-line influence upon these birds. Since this species apparently prefers to glide on updrafts along the ridge, one would expect fairly similar Red-tail counts at locations such as Bake Oven Knob and Hawk Mountain—particularly on northerly or northwesterly winds.

EXAMPLE NO. 1.—On 6 November 1966, brisk west winds brought 82 Red-tailed Hawks to Bake Oven Knob; 170 were counted at the North Lookout at Hawk Mountain. Throughout the day most of the hawks passed high overhead. The marked discrepancy in the counts suggests that many new birds joined the ridge flight between the two locations. Presumably the new hawks—those not counted at the Knob—were subject to wind drift prior to joining the ridge flight.

EXAMPLE NO. 2.—The Red-tail flight of 12 November 1966 occurred on moderate to brisk northwest winds with 245 counted at Bake Oven Knob and 266 seen at Hawk Mountain's North Lookout. Apparently the birds were following the ridge southward at least as far as Hawk Mountain. The minor differences in the counts were caused by a slightly longer period of observation at Hawk Mountain plus a few hawks which may have drifted to the Kittatinny Ridge and joined the flight southwest of the Knob.

EXAMPLE NO. 3.—On 11 November 1967, moderate to brisk east to southeast winds occurred at the South Lookout at Bake Oven Knob. The Red-tail count was 136 compared with 132 seen at the South Lookout at Hawk Mountain. Despite the movement of hawks along the south slope of the ridge—a condition which may result in many birds leaving the ridge southwest of Bake Oven Knob— the pattern of the flight was remarkably similar at the two locations. Evidently the updrafts were sufficiently favorable to keep the hawks following the ridge. Winds from these quarters, but of much lower velocity, almost certainly would have resulted in the flight becoming widely dispersed.

EXAMPLE NO. 4.—The brisk northwest morning winds of 26 October 1968 shifted to the west during the afternoon. At Bake Oven Knob, 161 Red-tailed Hawks were counted, the birds flying quite high and far out over the valley north of the ridge. Downridge, at the North Lookout at Hawk Mountain, 162 Red-tails were tallied. Despite the nearly identical counts, the data show discrepancies in the patterns of the flights and suggest that a portion of the birds observed at each location were not seen at the opposite site. This is not surprising since telescopes were required to confirm the identification of many birds due to poor light conditions and the considerable distance at which the hawks were flying north of the ridge.

EXAMPLE NO. 5.—On 30 October 1968, excellent Red-tail flights were seen both at Bake Oven Knob, where 350 were counted, and at Hawk Mountain's North Lookout where 436 were seen. Moderate west to northwest winds prevailed during the duration of the flight. Although the general patterns of the flights are similiar, it is evident that some new birds joined the ridge flight southwest of the Knob, thus resulting in the higher Hawk Mountain count. Once again, one must assume that the additional birds were experiencing wind drift prior to joining the ridge flight.

The previous examples, plus my general impression of Red-tail flights over the years, suggest that the Kittatinny Ridge between Bake Oven Knob and Hawk Mountain is an important flyway or diversion-line for these large buteos—particularly when northwest winds occur. Not uncommonly, larger flights seem to occur at Hawk Mountain than at the Knob. It was with considerable surprise, therefore, that the 1971 field studies conducted at three major Kittatinny Ridge lookouts mentioned earlier suggested that Red-tailed Hawks tend to decrease in numbers as they progress southwestward between Raccoon Ridge, New Jersey, and Hawk Mountain, Pennsylvania (Heintzelman, 1972f). However, discrepancies occurred in some flights, and the situation may have been atypical as 1972 data reported by Tilly (1973), by Heintzelman and MacClay (1973), and by Sharadin (1973) do not seem to verify the existence of such a trend.

Although Sharp-shinned, Red-tailed, and Broad-winged Hawks are the most abundant autumn migrants to be observed along the Kittatinny Ridge, migrant Ospreys also occur in moderate numbers. Moreover, they readily use updrafts along the ridge to aid them in their southward migrations. Hence they are good subjects to consider when evaluating the diversion-line influence of the ridge. Wind direction seems to exert a marked influence upon Osprey flights, however. For example, it has long been common knowledge among regular visitors to the eastern Pennsylvania hawk lookouts that Ospreys tend to stick closely to the ridge on northerly or northwesterly winds. In contrast, when easterly or southerly winds occur, these hawks frequently tend to break away from the ridge near the vicinity of Tri-County Corners, about midway between Bake Oven Knob and Hawk Mountain, and head toward the Pinnacle south of Hawk Mountain. Apparently the birds then follow this spur of the ridge southwestward for varying distances.

In addition to wind, and its influence upon Osprey migrations, the Little Schuylkill River flows perpendicular, then parallel, to the base of the north slope of Hawk Mountain. It is difficult to assess the potential influence of this type of geographic feature upon an aquatic raptor such as the Osprey, but it may act as another diversion-line for birds flying north of Hawk Mountain, drawing them to the Kittatinny Ridge. In any event, it seems worthwhile to examine some Osprey flights as seen from Bake Oven Knob and Hawk Mountain, Pennsylvania.

EXAMPLE NO. 1.—On 11 September 1965, four Pennsylvania lookouts were in use along the Kittatinny Ridge: Tott's Gap, Bake Oven Knob, Bear Rocks and Hawk Mountain. A northwest wind at 5 to 15 miles per hour, coupled with excellent visibility, prevailed throughout the day. At Bake Oven Knob I counted 102 migrating Ospreys, 40 appearing between 1500 and 1700 hours (Heintzelman, 1966: 33). This extraordinary count still remains the highest daily Osprey count known for the Kittatinny Ridge. Curiously, some marked discrepancies developed in the counts from the various lookouts. At Tott's Gap, for example, the pattern of the Osprey flight varied considerably from that noted at the Knob and Bear Rocks but recent studies at Tott's Gap suggest that this site frequently has limited value as a lookout (Heintzelman, 1973). In contrast, the Bake Oven Knob and Bear Rocks counts were nearly identical, as would be expected, the essential difference being due to a termination of observations at the latter site earlier than at the Knob. Downridge, at Hawk Mountain, the pattern of the Osprey flight showed relatively little similarity to that noted at Bake Oven Knob and Bear Rocks, however. It is difficult to explain the discrepancy, but the most likely explanation is that some birds were overlooked by observers at Hawk Mountain. Nonetheless, the possibility cannot be excluded that some birds drifted from the ridge after passing Bear Rocks.

EXAMPLE NO. 2.—Light east winds occurred on 12 September 1966 at Bake Oven Knob where 27 Ospreys were counted compared with 17 counted at Hawk Mountain's North Lookout. The general pattern for the flight, as recorded at each lookout, is fairly similar, however. Apparently some birds cut away from the ridge between the two sites as would be expected on easterly winds.

EXAMPLE NO. 3.—On 24 September 1966, moderately brisk to brisk west winds prevailed at Bake Oven Knob where 69 Ospreys were counted. The bulk of these birds passed at eye level, or a little lower, and not too far out over the north slope of the ridge. In contrast, 48 Ospreys were seen at the North Lookout at Hawk Mountain. Despite the fact that the birds were easy to see as they flew along the north slope at the Knob, a considerable discrepancy developed in the pattern of the Osprey flight at Hawk Mountain. In this case one is left with no alternative other than to conclude that as many as 30 percent of the Ospreys seen at Bake Oven Knob drifted from the ridge before reaching Hawk Mountain.

EXAMPLE NO. 4.—Light to moderate east to south winds occurred on 20 September 1967 at Bake Oven Knob where 46 Ospreys were seen compared with 33 at Hawk Mountain's North Lookout. Hence about 28 percent of the birds seen at the Knob failed to reach Hawk Mountain as might be expected on winds from easterly and southerly quarters.

EXAMPLE NO. 5.—Calm to light southeast to south winds, turning to the southwest during the afternoon, occurred at Bake Oven Knob on 24 September 1968. The Osprey count was 43 compared with 30 seen at Hawk Mountain's South Lookout. Hence about 30 percent of the Ospreys seen at the Knob again failed to reach Hawk Mountain on these winds—presumably because they drifted from the ridge after passing the Knob.

One of the difficulties in studying Sharp-shin, Red-tail, Broadwing, and Osprey flights is that individual birds cannot be recognized easily. Hence more readily recognizable individuals of rarer species sometimes offer better opportunities to evaluate a geographic feature's influence as a diversion-line. Bald Eagles are particularly well suited for this purpose because they can be separated into age categories (adult and immature) with ease. For this reason it has been customary to record the age and time of appearance of each eagle seen at Bake Oven Knob and Hawk Mountain, thereby making it possible to trace the movements of many individual birds between the two lookouts just as Frey (1940) did between Hawk Mountain and Sterrett's Gap, Pennsylvania.

During the period 1961 through 1967, for example, a careful comparison of the ages and times of appearance of Bald Eagles at Bake Oven Knob and Hawk Mountain, Pennsylvania, revealed many discrepancies in that about 51 (53.6 percent) of the 95 eagles observed at the Knob were not seen downridge at Hawk Mountain (table 57). This demonstrates that a much larger number of Bald Eagles migrate through the Bake Oven Knob-Hawk Mountain area than is reflected by eagle counts from either location.

By critically comparing these eagle records, the preparation of a composite Bald Eagle count (table 57) is possible. A few birds, included in the Hawk Mountain records but not actually seen there, were eliminated from the sample.

South of the Kittatinny Ridge many mountain ridges have been investigated as hawk lookouts. For example, Robbins (1950) coordinated an extensive one-day hawk watch in 1949 on most of the parallel ridges in Frederick, Washington, Allegany, and eastern Garrett counties, Maryland. This effort was aimed primarily at studying Broad-winged Hawk movements, but other Maryland studies demonstrate that Monument Knob is one of the state's best lookouts (Beaton, 1951; Carlson, 1966). Hawks drifting southwestward from Waggoner's Gap, Pennsylvania, may form the origin of a portion of the Monument Knob flights.

In Virginia, efforts have been made to study hawk flights along many ridges, including the Blue Ridge Parkway, but much systematic field study still is needed in this area before adequate conclusions can be reached regarding the importance of the various mountain peaks and fire towers as hawk lookouts and the diversion-line roles of these sites. However, in southwestern Virginia the efforts of the Tennessee Ornithological Society at the Mendota Fire Tower (Finucane, 1956–1972) demonstrate that this is one of the better spots in the southern Appalachians for observing Broad-wing flights.

Similarly, DeGarmo (1953) devoted much field study since 1947 to hawk migrations in West Virginia with particular emphasis upon Broad-wing migrations, and Hurley (1970) and Shreve (1970) provide more recent information on West Virginia hawk flights. Most of these migrations are detected along long mountain ridges (DeGarmo, 1953: 40), thus suggesting a definite diversion-line phenomenon, with a surprising number appearing on southeasterly winds as, for example, on the Allegheny front. Among West Virginia's more important lookouts are Bear Rocks (DeGarmo, 1953: 41; Hall, 1964), the Hanging Rocks Fire Tower (Hurley, 1970), and Middle Ridge (Shreve, 1970).

South of Virginia and West Virginia, as in Tennessee, a coordinated effort has been devoted to Broad-winged Hawk migrations for many years (Behrend, 1951–1954; Finucane, 1956–1972). These field studies demonstrate that scattered broad-front migrations cross Tennessee although some of the birds apparently use various mountain ridges for limited distances as diversion-lines if proper conditions prevail. South of Tennessee, only scattered hawk counts have been conducted.

OTHER AREAS

Aside from Mount Tom, Massachusetts, most inland lookouts in northern New England do not appear to be on major migration routes. However, data are still very incomplete from this area. Inland in southern New England the short-term field studies coordinated by Hopkins and Mersereau (1971, 1972) are providing preliminary coverage at many potential sites in western Massachusetts and Connecticut. These observers suggest that Broad-winged Hawks migrate in a definite wave having dimensions of length, width, and crest (Hopkins and Mersereau, 1971: 3). The suggested width of the wave covers about ten miles at its densest portion. This novel idea requires much additional field study to verify or reject its validity. During 1972 another short-term hawk watch again was coordinated in southern New England, and a wave-like migratory movement of Broad-winged Hawks again was suggested (Hopkins and Mersereau, 1972). These suggested waves are not the traditional concepts of avian migratory-wave phenomena as studied by Leck (1972), for example, for passerines appearing at a coastal New Jersey site during autumn. Leck's waves were correlated with the movements of weather systems such as cold fronts. Similar periodic waves of White-breasted Nuthatches were recorded in 1968 at Bake Oven Knob, Pennsylvania, by Heintzelman and MacClay (1971). Apparently the wave movements suggested by Hopkins and Mersereau (1971, 1972) occur within a restricted geographic area on a particular day but are not the result of the passage of weather fronts acting as releasers for large numbers of birds to begin migratory movements.

South of New England, hawk counts have been made from various nonmajor inland sites. In New York, for example, field studies at Hook Mountain show that a surprising number of hawks migrate along this relatively short east-west ridge (Mills and Mills, 1971; Thomas, 1971a, 1971b). Sharp-shinned, Broad-winged, and Sparrow Hawks appear in largest numbers, but Ospreys also are fairly numerous. Like Raccoon Ridge, New Jersey, Hook Mountain is particularly interesting since both the ridge and the Hudson River nearby might serve as diversion-lines.

Once past Hook Mountain, migrating hawks may use various minor ridges and ultimately join birds flying along New Jersey's Watchung Mountains. Eynon (1941) and Lang (1943) point out that the flights observed at the Montclair Hawk Lookout Sanctuary are a portion of the Watchung hawk migrations and suggest that

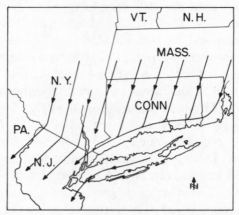

MAP 44   Probable routes followed by migrating hawks in autumn through Connecticut, southern New York, and northern New Jersey. Redrawn from Breck (1960b).

some of these birds follow a south-southeast course along rivers, although Eynon also thought that these hawks came from Connecticut. Hence Hook Mountain may funnel some hawks into the Watchungs via the Hudson River. Broad-front migrations (Lang, 1943: 349) also contribute to these Watchung flights.

Mount Peter is another of southeastern New York's hawk lookouts (Bailey, 1967, 1969; Rogers, 1971). Its position on Bellvale Mountain, just north of the New Jersey-New York border and well south of (but within sight of) the main fold of the Kittatinny Ridge, places it in an intermediate position between the major Atlantic coastline and Kittatinny Ridge migration routes and thus isolates it from a substantial portion of the hawk flights which usually are diverted for varying distances along portions of the Kittatinny Ridge. Although a portion of the Kittatinny flights sometimes drift south to form part of the Mount Peter flights, many birds also approach on a broad front from southern New England. In any event, Eynon (1941: 114) states that hawks enter New Jersey's Highlands region from a route southwest of Mount Peter. These are scattered flights, ultimately seen at Breakneck Mountain, Stag Lake, and the Bowling Green fire tower. The fire tower on Bearfort Mountain is another example of a site receiving hawks from Mount Peter and from scattered, broad-front migrations cross-

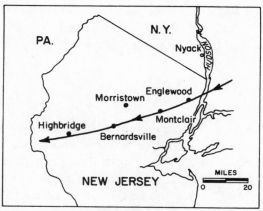

MAP 45   Probable routes followed by migrating hawks in autumn along the Watchung Mountains in northern New Jersey. Redrawn from Lang (1943).

ing the area between the Kittatinny Ridge and the Atlantic coast-line (Koebel, 1970).

In Pennsylvania, most hawk lookouts are located on the crest of the Kittatinny Ridge between Delaware Water Gap and Wag-goner's Gap above Carlisle. However, there are some secondary observation sites located north, south, and west of the main Kit-tatinny flyway which also produce some hawk flights of more modest size. For example, the modest flights observed at Pipersville by Ann Webster (personal communication), a site considerably south of the Kittatinny Ridge, apparently are composed mainly of hawks drifting cross-country, although some Ospreys observed at this location may have followed the Delaware River southward as a diversion-line before appearing over Pipersville. Similarly, two large kettles of Broad-winged Hawks seen on 5 October 1969 (an unusually late date) at Kutztown, well south of the Kittatinny Ridge (Nagy, 1970), apparently were drifting cross-country, too.

West and/or north of the Kittatinny Ridge, hawk flights of modest size occur along most of the ridges. The flights observed at the Pulpit on Tuscarora Mountain (Carl L. Garner, personal communications) are typical of such migrations and demonstrate that these ridges, although less important than the Kittatinny, none-theless act as diversion-lines for some migrant raptors. Hawks ob-served at the Pulpit, for example, never become part of the main

Kittatinny Ridge flights. Instead they use Tuscarora Mountain as another diversion-line, following it southward, and eventually join various hawk migrations crossing Maryland, Virginia, and West Virginia.

GREAT LAKES

The autumn hawk flights which occur at various concentration points along the northern and/or western shorelines of the Great Lakes are using the third major North American raptor migration route. In point of fact, several subroutes are involved in these migrations (Pettingill, 1962). At each of the Great Lakes their shorelines act as major diversion-lines, diverting birds around the open waters since many species dislike crossing large bodies of water. For example, Pettingill (1962) states that Red-tailed, Red-shouldered, Broad-winged, Rough-legged, and Marsh Hawks adhere strictly to around-the-water migration routes because the thermals and/or

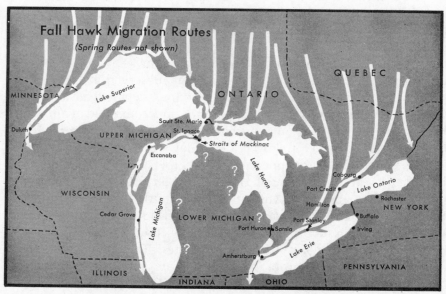

MAP 46 General autumn routes followed by migrating hawks in the Great Lakes region. Most hawks, particularly Broad-wings, tend to avoid crossing large expanses of open water. Reprinted from Pettingill (1962) in *Audubon*, the magazine of the National Audubon Society (copyright 1962).

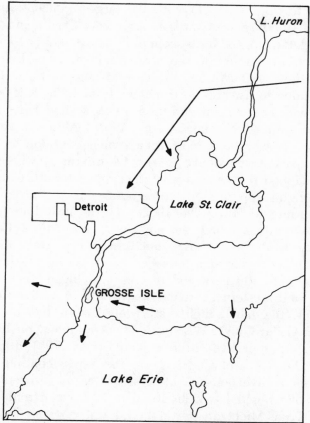

MAP 47 Routes followed by migrating Broad-winged Hawks during the autumn of 1953 in the Detroit, Mich., region. Redrawn from Merriam (1953).

strong air currents they require for their migrations are not found over large bodies of water. However, accipitrine hawks, eagles, Ospreys, and falcons will migrate over water if necessary (Perkins, 1964). Kleiman (1966) also observed Rough-legged Hawks migrating over Lake Erie after leaving Point Pelee. Similarly, Burns (1911) reported that some Broad-winged Hawks cross the southwestern end of Lake Erie in sight of a chain of islands which extend to the Ohio shore.

The hawk flights which occur along the north shoreline of Lake Ontario apparently drain from southern Quebec and eastern Ontario (Pettingill, 1962: 44) with the largest concentrations being seen from Cobourg westward through Port Credit and Hamilton. From the latter site the birds seem to fly cross-country toward Hawk Cliff near Port Stanley on the north shore of Lake Erie. Sizeable concentrations of hawks occur here (Haugh, 1972). The flights then continue to use the north shoreline of Lake Erie as a diversion-line with notable concentrations being seen at Point Pelee near the western end of the lake (Gray, 1961). Although some hawks may cross the lake here, most Broad-wings continue to follow the shoreline past Amherstburg to the Detroit area, where Merriam (1953) mapped the general migration routes. South of the Detroit area the flights apparently become scattered.

According to Pettingill (1962: 44) the next important Great Lakes hawk flights follow the northwest shoreline of Lake Huron toward the Sault Saint Marie-Saint Ignace area. Data still are very incomplete from this area, however. Presumably the hawks move overland to the northern end of Lake Michigan, then follow the northern and western shoreline southward as a diversion-line. In any event, the hawk flights are well known by the time they appear at Cedar Grove, Wisconsin, about midpoint on Lake Michigan's western shoreline (Mueller and Berger, 1961, 1967a, 1967b, 1968, 1973). South of Cedar Grove the hawks continue to follow the shoreline diversion-line at least as far as Milwaukee, where Jung (1935) noted hawk flights along the waterfront bluffs, but south of Lake Michigan the flights almost certainly become scattered.

That the shorelines of the Great Lakes act as diversion-lines for migrating hawks again is illustrated along the northern and western sides of Lake Superior. The source of these migrations appears to be central Ontario (Pettingill, 1962: 44). Particularly impressive concentrations of migrating hawks are best known at the southwestern tip of Lake Superior at Duluth, where the birds fly along high bluffs along the lake's shoreline (Green, 1962; Hofslund, 1954, 1962, 1966). Elsewhere in Minnesota some Bald Eagle and hawk migrations occur along the Upper Mississippi River (Reese, 1973), but most of the hawks passing Duluth probably become scattered south of Lake Superior.

17

Hawk Counts as Indices to Raptor Population Trends

One of the most difficult tasks confronting wildlife biologists is the accurate determination of population levels of wild animals. Rarely is it possible to count all individuals of a species directly. The problem is particularly acute in respect to estimating population levels of birds of prey which, as breeding birds, are distributed over large geographic areas. Hence virtually no data are available providing accurate estimates of raptor population levels on a continental basis for long periods of time. However, some efforts have been made to estimate current regional population levels of certain rare or endangered species or subspecies of birds of prey. Some examples include field studies of California Condors (Koford, 1953; Miller, McMillan and McMillan, 1965; McMillan, 1968), various Peregrine Falcon populations (Hickey et al., 1969), and various other diurnal birds of prey (Hickey et al., 1969). In at least one instance the *entire* breeding raptor population also was determined on a geographic area of modest size (Craighead and Craighead, 1956).

If accurate data on raptor population levels covering large geographic areas are scarce, adequate data are even rarer on the factors which regulate raptor numbers (Lack, 1954) and the factors responsible for causing periodic cycles in some hawk populations

(Keith, 1963). Hence it is not surprising that autumn hawk counts gathered at a major concentration point such as Hawk Mountain, Pennsylvania, have been used as indices to trends in diurnal raptor population levels in northeastern North America (Spofford *in* Hickey, 1969: 323–332). Considered at face value, analysis of the Hawk Mountain data might appear to be a suitable index to estimating changes in hawk populations. Hackman and Henny (1971) even have attempted to use hawk counts from a minor Maryland observation point similarly.

In using the Hawk Mountain data, Spofford (*in* Hickey, 1969) assigned the categories of declining trend, increasing trend, and no discernible trend to the species he considered. Those apparently declining were Sharp-shinned and Cooper's Hawks, Bald Eagles, Golden Eagles, and Peregrine Falcons. Species with populations apparently increasing were Red-shouldered, Broad-winged, and Marsh Hawks, Ospreys, and Sparrow Hawks. No discernible trend could be detected in Goshawk and Red-tailed Hawk populations. Spofford noted that considerably fewer Red-tails have been counted at Hawk Mountain since World War II, however.

Despite these attempts to develop rough indices to raptor population levels, some hawk watchers have expressed serious reservations regarding the validity of estimating population trends from hawk migration data (Heintzelman and MacClay, 1972: 17; Nagy, 1972: 3; Taylor, 1970: 2–3)—at least data from a single concentration point. These objections are based upon various considerations. Certainly, for such indices to be valid, at least two major conditions yearly would have to remain constant: (1) hawks observed during migration, e.g., at Hawk Mountain, yearly must originate from the same geographic area, and (2) the same percentage of the hawk population yearly must pass, and be detected, at the same concentration point. Neither condition can be established with certainty. Indeed, very little is known regarding the precise geographic areas from which migrant hawks come that are seen at Hawk Mountain and most other eastern lookouts. Moreover, it seems likely that the second condition seldom prevails from year to year. Numerous variables also can influence the percentage of a hawk population which yearly passes a lookout. Among these are raptor morphology and anatomy, wind and other weather conditions, noon lulls in daily flights, geography, and observer techniques. Endless combinations of these and other variables determine when, where, and how many migrant hawks are seen during autumn. Hence, in my opinion, the greatest caution must be exercised in considering northeastern raptor population indices based upon autumn hawk counts. Some specific comments regarding certain species follow.

GOSHAWK

Spofford's (*in* Hickey, 1969) analysis of Goshawk counts from Hawk Mountain, Pennsylvania, suggest no discernible long-range population trend. However, he points out that this species occasionally displays southward invasions or irruptive migratory movements of large proportions. Such an irruption occurred in the autumn of 1935 at Hawk Mountain (Broun, 1949a: 150). An invasion of even larger magnitude failed to occur again until the autumn of 1972, when enormous southward flights of Goshawks were observed at many widely separated locations in eastern North America. For example, 428 were counted at Hawk Mountain (Sharadin, 1973), 343 appeared at Bake Oven Knob (Heintzelman and MacClay, 1973), and 202 were seen at Raccoon Ridge, New Jersey (Tilly, 1973). However, the main thrust of the invasion occurred along the western Great Lakes, where observers at Duluth, Minnesota, counted 5,152 Goshawks (P. B. Hofslund, personal communication).

In view of the extraordinary 1972 Goshawk invasion, raptor biologists seemingly have little choice but to revise their concepts of Goshawk population levels in eastern North America. However, extreme caution must be taken in predicting that Goshawk populations will remain at such previously unsuspected high levels. In point of fact, as frequently occurs with other cyclic predator-prey phenomena (Keith, 1963), it is likely that Goshawk populations may "crash" and remain at relatively low levels before gradually building to a cyclic high again. However, for such a massive eastern invasion to again occur, a crash of all principal Goshawk prey populations presumably would have to occur synchronously as it apparently did prior to the 1972 invasion. In addition, the Goshawk population may have to suffer an extremely unproductive breeding season. Apparently this rarely occurs. More commonly, cyclic prey tend to exhibit nonsynchronous cycles in different parts of their range. Under such conditions Goshawks might undergo irruptive migrations only in certain geographic areas. Much remains to be learned about Goshawk populations, and their occasional invasions southward, however.

BROAD-WINGED HAWK

In eastern North America, most hawk watchers have placed particular attention upon one species—the Broad-winged Hawk. Ironically, of all eastern hawks, the Broad-wing is the least desirable species one could select for studying hawk migrations. For example, it is extremely atypical in its flight pattern because it is the only eastern hawk which makes extensive use of thermals. Sometimes hundreds of these birds flock within a single rising bubble of warm air. Moreover, it seems much more prone to undergoing wind drift (a condition related to its thermal dependency) than most other diurnal birds of prey. Additionally, it is extremely susceptible to exhibiting noon lulls in its mid-September flights, which can seriously alter the accuracy of Broad-wing counts. Thus it is extremely unlikely that Broad-winged Hawks satisfy either of the two major conditions in their autumnal migrations.

Therefore, in view of the migratory characteristics of this species, it appears likely that the suggested increasing trend mentioned by Spofford (*in* Hickey, 1969) for Broad-wing populations in the northeastern section of North America may be invalid.

ROUGH-LEGGED HAWK

Although Spofford did not consider the Rough-legged Hawk, some comments on its autumn migrations nonetheless may be useful. Rough-legs are among the rarest migrants observed at Bake Oven Knob and Hawk Mountain, Pennsylvania (Heintzelman, 1969b; Broun, 1949a). During some years none appear. In other years only a handful are counted. However, the species is known to be cyclic. Hence, during years when Rough-legs reach exceptionally high population levels, an unusually large number may appear in the late season hawk flights at some concentration points.

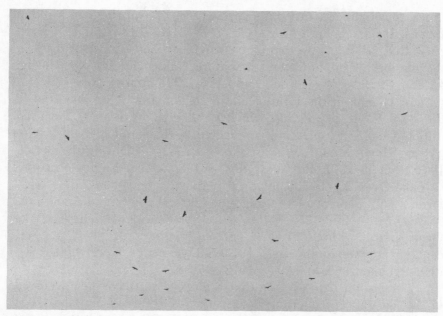

A kettle of Broad-winged Hawks inside a thermal. These birds frequently are subject to wind drift. Photo by Donald S. Heintzelman.

This apparently happened in 1971 at Bake Oven Knob, where 37 Rough-legged Hawks were counted (Heintzelman and MacClay, 1972). In contrast, only 5 were counted there in 1972 (Heintzelman and MacClay, 1973). As with Goshawks, autumn hawk counts presumably are valid as reflectors of Rough-leg populations only when this species exhibits cyclic peaks.

GOLDEN EAGLE

Golden Eagles from the small northeastern Appalachian population are rare, but regular autumn migrants at Bake Oven Knob and Hawk Mountain, Pennsylvania (Heintzelman, 1969b; Broun, 1949a: 187–195). Based upon studies of eagle counts at Hawk Mountain (Spofford *in* Hickey, 1969) and breeding bird surveys (Spofford, 1971), a decline in the numbers of northeastern Golden Eagles has been postulated. Spofford (1971: 3–7) attributed the suggested population decline to a DDT-caused thin-eggshell syn-

drome occurring in breeding populations of Appalachian Golden
Eagles. Heintzelman and MacClay (1972: 18–19) point out, however, that "If this were true, one would expect a marked drop in
the number of sub-adult and immature Golden Eagles seen during
autumn at Bake Oven Knob (and Hawk Mountain). Precisely this
type of change in the ratio of adults to immatures has occurred in
the Bald Eagle population which is affected by the DDT-caused
thin eggshell syndrome and is reflected in the count of Bald Eagles
passing Hawk Mountain and Bake Oven Knob." During the period
1961 through 1971, the age ratios of 193 Golden Eagles observed
at Bake Oven Knob did not drop markedly, however. Adult birds
formed 63.2 percent of the migrants and subadults and immatures
36.8 percent (table 58). In 1972, 42 Golden Eagles passed Bake
Oven Knob of which 20 (47.6 percent) were adults, 12 (28.6 percent) were subadults, and 10 (23.8 percent) were immatures
(Heintzelman and MacClay, 1973).

These data demonstrate that a marked shift has not occurred
in the age ratio of migrant Golden Eagles observed at Bake Oven
Knob, Pennsylvania. Heintzelman and MacClay (1972: 19) point
out, however, that "A pesticide caused breeding failure of Golden
Eagles nesting in the Appalachians presumably would cause a
marked shift in the age composition of the population of these birds
just as it has in the Bald Eagle population. Autumn counts of migrating Golden Eagles passing Bake Oven Knob and Hawk Mountain presumably would reflect this shift more readily than a change
in the actual population level of the species." Comparisons of the
ages and exact times of appearance of various Golden Eagles at
Bake Oven Knob and Hawk Mountain demonstrate that many
more eagles occur in the area than are observed at either observation point (Broun, 1966: 16). Frey (1940) reached similar conclusions after comparing Golden Eagle records from Sterrett's Gap,
Pennsylvania, with those from Hawk Mountain.

BALD EAGLE

Bald Eagles also are autumn migrants along the Kittatinny
Ridge (Heintzelman, 1969b; Broun, 1949a). In recent years much
of the continental population of this species in the contiguous states

has declined to alarmingly low levels (Sprunt *in* Hickey, 1969: 347–351), and the southern Bald Eagle now is included on the federal list of endangered wildlife. Various causes doubtless have contributed to the southern Bald Eagle's population decline—among them human disturbance, habitat loss, and shooting (Sprunt *in* Hickey, 1969)—but the most serious cause responsible for the rapid post-World War II decline is the adverse affect which DDT has upon the reproductive physiology of these birds, resulting in the now well-known thin-eggshell syndrome currently affecting many raptor species.

Because the southern Bald Eagle population has declined so drastically since World War II, the effect presumably could not avoid being reflected in Bald Eagle counts at Hawk Mountain, Pennsylvania, since Broley (1947) demonstrated that Florida eagles migrate past this site during autumn; Spofford's (*in* Hickey, 1969) analysis of these counts reflect this decline. However, it must be remembered, as demonstrated earlier in table 57, that many more Bald Eagles migrate along sections of the Kittatinny Ridge than are counted at any single location.

Perhaps more significant, therefore, in terms of evaluating the usefulness of counts of migrants, is the marked change in the age ratios of Bald Eagles seen at Bake Oven Knob and Hawk Mountain. For example, Broun (1961: 9) reported that 36.5 percent of the Bald Eagles counted at Hawk Mountain during the years 1931 through 1941 were immature, whereas 23.1 percent were immature between 1954 and 1960. Similarly, at Bake Oven Knob, Heintzelman and MacClay (1972: 20) report that only 33 (18.6 percent) of the 177 Bald Eagles counted during the period 1961 through 1971 were immatures (Table 59). This contrasts sharply with the 36.8 percent figure for immature and subadult Golden Eagles seen at the Knob during the same period.

OSPREY

One of the most perplexing discrepancies in using hawk counts as population indices is illustrated by recent Osprey counts from Hawk Mountain and elsewhere. Despite widespread agreement that eastern Osprey populations have suffered disastrous reproductive failures during recent years, enormous numbers of migrant Ospreys recently have been seen at Hawk Mountain. Similarly, high Osprey counts also have been reported at Bake Oven Knob, Pennsylvania (Heintzelman and MacClay, 1972, 1973). Hence it is not surprising that Spofford's (*in* Hickey, 1969) analysis of Hawk Mountain data suggest a population increase for Ospreys, although it is extremely unlikely that such a population increase has, in fact, occurred.

There are various reasons why higher Osprey counts have been recorded at Hawk Mountain in recent years. Among them are more observers, longer periods of observation daily, the use of two lookouts and, doubtless, some duplication of counts. Nonetheless, despite all of these variables combined, one still is faced with the fact that unusually large numbers of Ospreys are migrating southward along the inland ridges at a time when an increasingly large proportion of the eastern Osprey population continues to experience serious reproductive failure.

Why, then, are more Ospreys counted during migration when fewer and fewer nestlings are being fledged on the breeding grounds? There are no completely acceptable answers to this fascinating problem. However, several speculative explanations have been advanced. Peterson (1966: 10), for example, stated that "I suspect that these birds are coming from unpolluted Canadian lakes. They are passing rapidly down the ridges to their wintering grounds in the Caribbean and elsewhere, where they are probably finding less competition than formerly from the Ospreys of our North American Coast."

Another speculative explanation (Heintzelman, 1970c: 122) focused on the possibility that Ospreys recently may have changed their migration patterns or routes due to DDT and/or its metabolites. "Is it possible that the birds have slightly altered the normal physiology of energy production in their major flight muscles, thereby resulting in somewhat less than normal amounts of energy being produced? If so, migrating Ospreys—which may not have been greatly dependent upon strong air currents along mountain ridges such as Hawk Mountain during years prior to the use of DDT —may be dependent to a greater extent upon the strong mountain air currents which permit the birds to glide relatively effortlessly

Osprey populations in many parts of the United States have experienced serious reproductive failures during recent years due to pesticide pollution, destruction and loss of habitat, and other factors. Hence nestlings such as this bird are becoming scarce along coastal New Jersey, and elsewhere, despite increasingly large numbers of migrating Ospreys being reported from some northeastern hawk lookouts. Currently there are no adequate explanations for these inconsistencies. Photo by Donald S. Heintzelman.

and with a minimum of energy being exerted. An unusual dependence upon mountain air currents could result in increased numbers of Ospreys at Hawk Mountain.

"Perhaps there is an indirect or direct physiological effect of DDT upon energy production in Osprey flight muscles."

Taylor (1971: 3), in suggesting another explanation similar to Peterson's, speculated that inland Osprey populations ". . . must be steadily increasing, possibly due to less wintering ground competition from the DDT-ridden, dwindling coastal breeding birds."

In point of fact, no completely adequate explanation currently is available to account for the discrepancy between the extremely low reproductive success of eastern Ospreys and current high Osprey counts during migration. Hence one should exercise the greatest caution in equating migration counts with current Osprey populations.

PEREGRINE FALCON

The disastrous population crash of North American Peregrine Falcon populations, particularly the subspecies *anatum,* due to pesticide-induced reproductive failure, is now well known (Heintzelman, 1970e: 35–38). Currently all North American Peregrine Falcons are included on the federal list of endangered wildlife.

Peregrines never have been more than minor components of the inland ridge flights (Broun, 1949a; Heintzelman, 1969b) compared with the notable numbers reported at Cape May Point, New Jersey (Allen and Peterson, 1936), Assateague Island, Maryland (Berry, 1971; Ward and Berry, 1972), and elsewhere. Hence any attempted use of Hawk Mountain Peregrine counts seemingly would be relatively meaningless. However, current continental Peregrine populations are so depressed that Spofford's (*in* Hickey, 1969) analysis could not avoid reflecting the drastic decline of this species. Curiously, recent counts from Cape May Point, New Jersey (Choate and Tilly, 1973; Clark, 1972), fail to reflect the near total elimination of the Peregrine Falcon as a breeding bird in most of eastern North America. Perhaps many of these birds are from the tundra subspecies which is not yet extirpated as a breeding bird.

SPARROW HAWK

Sparrow Hawks are primarily coastal migrants although moderate numbers occur at inland concentration points such as Hawk Mountain and Bake Oven Knob, Pennsylvania (Broun, 1949a; Heintzelman, 1969b). According to Spofford (*in* Hickey, 1969) the Hawk Mountain counts show an upward population trend in this species in recent years. Despite the numerous variables which can influence such counts, and analyses of them, the extraordinary Sparrow Hawk flight (about 25,000 individuals) reported at Cape May Point, New Jersey, on 16 October 1970 (Choate, 1972) adds support to an upward trend in the eastern Sparrow Hawk population. More important this count, along with the extraordinary 1972 Goshawk invasion, again suggests the need for raptor biologists to examine their current views of continental hawk population levels.

Part 6
Evolution of the
Broad-Winged Hawk

18

Migration and Evolution
in the Broad-Winged Hawk

Of the hawks which migrate across eastern North America, none
is more intensively studied than the Broad-winged Hawk. And few
species present as complex a problem for researchers—complex in
terms of its migration patterns and their relationship to the evolution of the species. It is part of a natural complex of five New World
species which Johnson and Peeters (1963) refer to as woodland
buteos: the Roadside Hawk, Ridgway's Hawk, Red-shouldered
Hawk, Broad-winged Hawk, and Gray Hawk.

TAXONOMY

The Broad-wing is polytypic, currently separated into six subspecies, five of which are insular (Brown and Amadon, 1968; Friedmann, 1950; Hellmayr and Conover, 1949). In itself, that is of considerable interest since the mainland race apparently avoids crossing large bodies of water during migration. Nonetheless, James
Bond (letter of 1 December 1969) points out that "Isolation of
comparatively small populations doubtless has accelerated racial
variation in the Antilles." Map 48 illustrates the geographic distribution of the subspecies, and table 60 summarizes Friedmann's
(1950) measurements for each race.

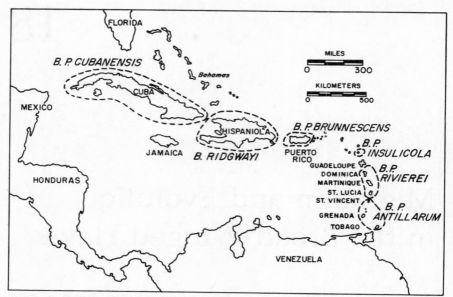

MAP 48   Geographic distribution of the insular subspecies of the Broad-winged Hawk in the West Indies. The distribution of Ridgway's Hawk also is shown.

Buteo platypterus platypterus (VIEILLOT)

This noninsular subspecies is highly migratory. It breeds in North America but winters in Florida, Central America, and South America (AOU, 1957). Recent winter records in North America north of Florida often are questionable, however.

Buteo platypterus cubanensis BURNS

This nonmigratory form is resident in Cuba and the Isle of Pines, and may have occurred in Hispaniola (Friedmann, 1950; Bond, 1956; Hellmayr and Conover, 1949). The inclusion of the Hispaniolan record is entirely on geographic grounds. Bond (1956: 29; letter of 1 December 1969) states that the record may represent the nominate race, a stray of either *cubanensis* or *brunnescens*, or the bird may have been collected elsewhere.

Adults (sexes alike) are similar to the immature plumage of the nominate race. The dark markings on the breast are streaks instead of true bars. Juvenile specimens (sexes alike) are indistinguishable from similar age specimens of the nominate race.

Buteo platypterus brunnescens DANFORTH AND SMYTH

This rare subspecies is known only from the type locality,* El Yunque Mountain, Puerto Rico, and from a few records at Utuado and Maricao, Puerto Rico (Friedmann, 1950; Bond, 1956; Leopold, 1963: 27). Hellmayr and Conover (1949: 116) suggest that an immature Broad-wing secured near Santiago, Dominican Republic, Hispaniola, may be referrable to this race. Neither Brown and Amadon (1968: 581) nor Bond (1956) comment on this.

This is the darkest of all subspecies of the Broad-winged Hawk. It is similar to *cubanensis* but is darker with the markings more blackish (Friedmann, 1950; Brown and Amadon, 1968).

Recently Recher and Recher (1966: 153–154) rejected the validity of *brunnescens*, advocated by Bond (1956, 1968), because the subspecies was described on the characteristics of a single bird. However, Bond (letter of 1 December 1969) points out that they did not see the type, and that "*B. p. brunnescens* is clearly a valid race." Nonetheless, Recher and Recher (1966) feel that their observations of Broad-winged Hawks in Puerto Rico during the spring of 1964 and 1965, as well as previous records from the island, represent migrating North American birds, i.e., *B. p. platypterus*. But Bond (letter of 1 December 1969) points out that "If North American *platypterus* migrates through the Antilles, it should have been recorded from such a well-known island as Jamaica." In view of the normal migration patterns of nominate race Broad-wings, it seems unlikely that a regular migration of mainland birds crosses the West Indies despite the recent collection of a nominate race bird on Cuba (James Bond, letter of 16 April 1973). A comparative series of recently collected Broad-wings from Puerto Rico will be required to determine if they represent resident *brunnescens* or another race.

* The geographic location where the first individual, or series of individuals, of the species or subspecies of animal (or plant) is collected.

Buteo platypterus insulicola RILEY

This form is confined chiefly to the southern portion of Antigua (Bond, 1956). It is the smallest and lightest colored of all of the races of the Broad-wing (Friedmann, 1950; Brown and Amadon, 1968). Measurements from only one bird are available, however.

Buteo platypterus rivierei VERRILL

This subspecies is resident on Dominica, Martinique, and Saint Lucia. It is a small form similar to *antillarum* but is slightly darker generally (Friedmann, 1950; Bond, 1956).

Buteo platypterus antillarum CLARK

This form is resident on Barbados (formerly?), Saint Vincent, the Grenadines (including Bequia, Mustique, Isle Quatre, Cannouan, Carriacou), Grenada, Tobago, and Little Tobago (Friedmann, 1950; Bond, 1956). Brown and Amadon (1968) consider it paler and slightly larger than *rivierei*.

MIGRATION ROUTES AND SPECIATION

Mainland Broad-winged Hawks are unusually dependent upon thermals to complete their migrations between North America and Central and South America. Nowhere do they normally cross large bodies of water. Instead they divert along shorelines or coastlines, using these geographic features as diversion-lines. Apparently the hawks are forced to do so because thermals are not formed over water due to its relatively homogeneous surface, which is unsuitable for thermal production.

Nonetheless, five of the subspecies of the Broad-wing are non-migratory insular forms. Apparently all are relatively recent in origin (James Bond, verbal communication). Moreover, these races apparently are derived from mainland birds.

How, then, can one explain the evolution of the insular subspecies? The answer appears to be keyed to the behavior of the bird which, in turn, is related to certain physical factors.

To begin, the aerodynamics and flight patterns of Broad-wings may have permitted some individuals or small groups to reach various islands in the Antilles accidentally. Perhaps this occurred on several occasions. Such a group of hawks may have been carried accidentally off course as storm waifs and thus arrived on one or more West Indian islands (Dean Amadon and Eugene Eisenmann, verbal communications). Once there they would tend to be power-less to leave because of the lack of thermal activity over the waters surrounding the island. Hence an isolated gene pool resulted and subspeciation occurred.

To consider this hypothesis further, I examined the geography of the West Indies in more detail. For example, the following distances exist between the closest mainland points and the various Antillean islands, with distances measured in statute miles.

Florida to Cuba (mainland)	125 miles
Cuba to Hispaniola	53 miles
Hispaniola to Puerto Rico	72 miles
Antigua to Dominica	96 miles
Dominica to Martinique	25 miles
Martinique to Saint Lucia	20 miles
Saint Vincent to Grenada	68 miles
Grenada to Tobago	82 miles

To further test the hypothesis that Broad-wings can ride ther-mals aloft but cannot glide over open waters for distances sufficient to travel from island to island, one must determine the distances covered by Broad-wings in interthermal glides. Unfortunately few data of this type are available. Stearns (1948b, 1949) observed Broad-wings gliding from a thermal at 2,000 feet over land in New Jersey. These birds covered a distance of four miles with a glide to fall ratio of about 12 to 1. Murton and Wright (1968) also reported these hawks reaching altitudes of 10,000 feet in thermals. A few other measurements also are available. Based upon the New Jersey data, it would appear impossible for Broad-wings to make inter-island flights successfully. However, if the data which Murton and Wright report from the Great Lakes are typical, interisland flights might be possible under some circumstances. Nonetheless, most available data suggest that interisland flights occur rarely or at very irregular intervals and then due to storms carrying the birds to other islands accidentally.

CONFLICTING DATA

Some data conflict with most Broad-wing migration information, and the previous hypothesis, however. For example, Robertson and Ogden (1968: 27), commenting on the migration of Broad-wings over the Florida Keys during the autumn of 1967, report: "As usual, the Broad-wing was by far the most common migrant, and the action was in the Keys, where thousands passed over Key West in the middle two weeks of October. The largest group reported anywhere else was a paltry 11 at Pine Is., October 22 . . . . About 1 p.m., October 11, at Key West, Mrs. Hames watched 'a huge, barrel-shaped formation' of Broad-wings, head south-southwest and hold that course until out of sight. An hour later, came two more flocks of 1000 each that merged as they gained altitude over the town and left on the same heading. Others at mid-day on the 16th, and again on the 20th, repeated the performance in exact detail. On the 19th, Mrs. Bonney found 'great flocks' that included a number of adults, fighting head winds over Long Key, 50 miles northeast of Key West. The season's last big flight reached Key West under overcast skies, mid-morning of the 21st. Mrs. Hames reported that this flock seemed disoriented and milled aimlessly for nearly an hour before it circled to altitude and left the area, also going south-southwest."

"The fall flights of soaring hawks in south Florida deserve much more attention than they have received. It would be interesting to know how these masses of Broad-wings reach the Keys without being detected more often in the peninsula. It is also extraordinary for soaring birds to undertake wide water crossings, as the flocks leaving Key West this fall apparently did. Their closest landfall as they headed was the north coast of Cuba, 100 miles away, and the least-water route for continuing fall flight beyond Cuba would involve a 125-mile crossing of the Yucatan Channel."

"If the flights do go south from Key West, it is remarkable that

Swainson's Hawk and the North American races of the Broad-wing are unknown or almost so in Cuba, Yucatan, and British Honduras. If they don't continue south, what becomes of them?"

A clue to the fate of these birds may be hidden in the fact that the last Broad-wing flights leaving Key West did so under overcast skies. This suggests that the birds were forced to lower altitudes because of a reduction in thermal activity. The fate of these hawks remains unknown, but radar studies at Key West might be helpful in solving this problem. It seems most likely that the birds returned to the Keys or to the mainland.

An alternative possibility is that the hawks continue flying as far as possible over the open waters and eventually perish in the sea. The anatomy and morphology of this species seem inadequate to permit the birds to carry out vigorous flapping flight for 100 miles or more in order to reach Cuba. Nonetheless, Robertson (1972) reported a fair number of Broad-wings over the Dry Tortugas, and Bond (letter of 16 April 1973) recently reported the collection of a nominate race Broad-wing in Cuba.

In any event, Robertson (*in* Bagg, 1970: 8) considers the most likely explanation for large numbers of Broad-winged Hawks appearing in Florida during autumn is that they are birds which have drifted off-course moving southeastward to Florida. Recently Tabb (1973) discovered a wintering population of Broad-wings in southern Florida, however. Elsewhere along the Gulf of Mexico nobody has observed a departure of Broad-wings over the Gulf (G. H. Lowery, Jr., letter of 3 January 1969) although one bird was recorded in spring over the open waters of the Gulf sixty miles from Louisiana (Lowery and Newman, 1954: 536). This bird probably was a storm waif.

Finally, one should not overlook the very early observations of the naturalist Oviedo, who observed hawks migrating across the West Indies between 1526 and 1535 (Baughman, 1947). His observations are not sufficiently detailed to permit one to determine with certainty the species involved in these flights, however. It is possible that he saw Ospreys or Peregrine Falcons, which regularly cross the Antilles. On the other hand, he may have witnessed one of the rare movements of Broad-wings involved in interisland flights.

RIDGWAY'S HAWK ON HISPANIOLA

A curious sidelight on the distribution of the Broad-winged Hawk in the Antilles is that this species is known on Hispaniola from a single specimen. Taking its place is another woodland buteo, Ridgway's Hawk, restricted to Hispaniola, including Grand and Petite Cayemite, Ile a Vache, La Gonave, and Beata. This species is common locally and apparently is most numerous on Grand Cayemite (Bond, 1956; Brown and Amadon, 1968). The latter authors suggest that the absence of the Broad-wing on Hispaniola may reflect a competitive reaction between it and Ridgway's Hawk—a suggestion with which James Bond agrees (letter of 1 December 1969).

Appendix 1
Scientific Names of Birds

The Thirty-Second Supplement to the American Ornithologists' Union Check-List of North American Birds (*Auk* 1973, 90: 411–419) contains a few changes in the vernacular and scientific names of birds of prey. Since these changes generally are not reflected in most of the hawk migration literature cited in this book, I have not used them. However, both the old and new names are included in this appendix.

Turkey Vulture	*Cathartes aura*
Black Vulture	*Coragyps atratus*
California Condor	*Gymnogyps californianus*
Goshawk	*Accipiter gentilis*
Sharp-shinned Hawk	*Accipiter striatus*
Cooper's Hawk	*Accipiter cooperii*
Red-tailed Hawk	*Buteo jamaicensis*
Red-shouldered Hawk	*Buteo lineatus*
Broad-winged Hawk	*Buteo platypterus*
Ridgway's Hawk	*Buteo ridgwayi*
Swainson's Hawk	*Buteo swainsoni*
Rough-legged Hawk	*Buteo lagopus*

Gray Hawk . . . .   *Buteo nitidus*
Roadside Hawk . . . .   *Buteo magnirostris*
Golden Eagle . . . .   *Aquila chrysaetos*
Bald Eagle . . . .   *Haliaeetus leucocephalus*
Marsh Hawk . . . .   *Circus cyaneus*
Osprey . . . .   *Pandion haliaetus*
Gyrfalcon . . . .   *Falco rusticolus*
Prairie Falcon . . . .   *Falco mexicanus*
Peregrine Falcon . . . .   *Falco peregrinus*
Merlin or Pigeon Hawk . . . .   *Falco columbarius*
Kestrel . . . .   *Falco tinnunculus*
American Kestrel or Sparrow Hawk .   *Falco sparverius*
Great Horned Owl . . . .   *Bubo virginianus*
Long-eared Owl . . . .   *Asio otus*
Saw-whet Owl . . . .   *Aegolius acadius*
White-breasted Nuthatch . . . .   *Sitta carolinensis*

Appendix 2
Tables

TABLE 1

HAWKS BANDED AT TRI-COUNTY CORNERS

Species	1959	1960	1961	1962	1963	1964	1965	1966	1967	1968	1969	1970
Goshawk	0	1	1	7	7	5	29	6	5	37	33	9
Sharp-shinned Hawk	0	0	1	64	76	73	146	89	70	105	112	154
Cooper's Hawk	3	0	1	9	6	4	21	14	16	16	12	9
Red-tailed Hawk	11	10	27	104	165	144	229	136	118	255	184	92
Red-shouldered Hawk	0	3	0	10	5	3	12	3	1	6	1	3
Broad-winged Hawk	2	1	2	16	11	25	22	8	6	7	1	3
Golden Eagle	0	1	0	1	0	0	1	2	1	3	0	1
Marsh Hawk	0	1	0	0	1	2	0	2	1	0	0	2
TOTALS	16	17	32	211	271	256	460	260	218	429	343	273

SOURCE: Compiled from data provided by C. J. Robertson.
NOTE: The number of observation days per year is provided in the tables whenever the information is available.

TABLE 2

HAWKS BANDED AT
LEHIGH FURNACE GAP-LEHIGH GAP

Species	1963	1964	1965	1966	1967	1968	1969
Goshawk	0	2	29	1	1	21	4
Sharp-shinned Hawk	6	0	0	0	0	0	0
Cooper's Hawk	0	0	3	3	0	2	2
Red-tailed Hawk	8	96	141	51	12	60	103
Red-shouldered Hawk	0	2	5	1	0	4	2
Broad-winged Hawk	0	0	1	1	0	0	0
Rough-legged Hawk	0	1	1	0	0	0	0
Golden Eagle	0	0	0	0	0	1	0
TOTALS	14	101	180	57	13	88	111

SOURCE: Compiled from data provided by John B. Holt, Jr.

TABLE 3

HAWKS BANDED
AT CAPE MAY POINT

Species	1967	1968	1969	1970	1971	1972
Goshawk	0	0	0	0	0	12
Sharp-shinned Hawk	12	31	112	52	356	1,025
Cooper's Hawk	2	7	7	10	9	29
Red-tailed Hawk	3	2	6	2	57	44
Red-shouldered Hawk	1	1	4	0	3	5
Broad-winged Hawk	2	0	2	2	5	1
Marsh Hawk	1	0	0	1	5	4
Peregrine Falcon	1	3	0	4	1	5
Pigeon Hawk	0	22	17	25	78	131
Sparrow Hawk	108	110	123	141	638	577
TOTALS	130	176	271	237	1,152	1,834*

SOURCE: Compiled from data published by W. S. Clark (1968, 1969, 1970, 1971, 1972, 1973).

* Includes one Kestrel (*Falco tinnunculus*).

TABLE 4

QUEBEC AUTUMN HAWK COUNTS (1970)

Species	September	October	November
Sharp-shinned Hawk	12	13	0
Cooper's Hawk	7	1	0
Red-tailed Hawk	4	5	1
Red-shouldered Hawk	2	4	0
Broad-winged Hawk	17	0	0
Rough-legged Hawk	2	5	5
Golden Eagle	1	0	0
Bald Eagle	0	0	1
Marsh Hawk	10	8	1
Osprey	13	1	0
Peregrine Falcon	5	0	0
Pigeon Hawk	3	0	1
Sparrow Hawk	24	9	4
TOTALS	100	46	13

SOURCE: Compiled from data in *Bulletin Ornithologique* (1970).

TABLE 5

MALDEN SCHOOL AUTUMN HAWK COUNTS

Species	1954	1956
Turkey Vulture	2	1
Sharp-shinned Hawk	30	22
Cooper's Hawk	8	0
Red-tailed Hawk	21	3
Red-shouldered Hawk	0	4
Broad-winged Hawk	10,116	2,756
Rough-legged Hawk	1	0
Bald Eagle	3	2
Marsh Hawk	24	68
Osprey	2	0
Sparrow Hawk	21	24
Unidentified hawks	866	200
TOTALS	11,094	3,080
No. Observation Days	8	3

SOURCE: Compiled from data filed by R. D. Merriam and Millie Reynolds with the U.S. Fish and Wildlife Service at Patuxent Wildlife Research Center.

TABLE 6

HAWK CLIFF AUTUMN HAWK COUNTS (1967)

Species	No. Counted
Turkey Vulture	202
Goshawk	17
Sharp-shinned Hawk	3,738
Cooper's Hawk	75
Red-tailed Hawk	4,795
Red-shouldered Hawk	1,208
Broad-winged Hawk	18,306
Rough-legged Hawk	55
Golden Eagle	8
Bald Eagle	6
Marsh Hawk	535
Osprey	65
Peregrine Falcon	18
Pigeon Hawk	8
Sparrow Hawk	2,346
Unidentified hawks	194
TOTAL	31,576

SOURCE: Compiled from Haugh (1972).

TABLE 7

PORT CREDIT AUTUMN HAWK COUNTS

Species	1953	1956	1957	1958	1959
Turkey Vulture	0	4	6	15	26
Goshawk	0	0	4	12	11
Sharp-shinned Hawk	23	41	72	61	55
Cooper's Hawk	2	29	25	30	33
Red-tailed Hawk	2	7	35	60	47
Red-shouldered Hawk	2	7	3	4	4
Broad-winged Hawk	7,991	5,809	1,671	4,883	2,519
Rough-legged Hawk	0	0	0	10	3
Golden Eagle	0	0	2	2	3
Bald Eagle	3	5	3	3	9
Marsh Hawk	12	12	45	51	49
Osprey	13	34	26	63	48
Gyrfalcon	0	0	1	4	5
Peregrine Falcon	0	3	4	11	10
Pigeon Hawk	4	0	3	17	28
Sparrow Hawk	9	51	50	49	116
Unidentified hawks	101	232	198	303	315
TOTALS	8,162	6,234	2,148	5,578	3,281
No. Observation Days	2	13	12	15	19

SOURCE: Compiled from data filed by Lucy McDougall with the U.S. Fish and Wildlife Service at Patuxent Wildlife Research Center.

TABLE 8
TORONTO AUTUMN HAWK COUNTS (1957)

Species	No. Counted
Turkey Vulture	5
Sharp-shinned Hawk	5
Cooper's Hawk	12
Red-tailed Hawk	19
Red-shouldered Hawk	2
Broad-winged Hawk	1,366
Bald Eagle	1
Marsh Hawk	12
Osprey	2
Pigeon Hawk	1
Sparrow Hawk	22
Unidentified hawks	159
TOTAL	1,606
No. Observation Days	8

SOURCE: Compiled from data filed by W. W. H. Gunn with the U.S. Fish and Wildlife Service at Patuxent Wildlife Research Center.

TABLE 9

DETROIT REGION AUTUMN HAWK COUNTS

Species	1951	1953	1954
Goshawk	1	0	0
Sharp-shinned Hawk	198	157	205
Cooper's Hawk	58	24	26
Red-tailed Hawk	5	7	25
Red-shouldered Hawk	28	10	5
Broad-winged Hawk	6,431	8,030	16,878
Rough-legged Hawk	4	17	1
Golden Eagle	1	0	0
Bald Eagle	10	1	6
Marsh Hawk	98	366	94
Osprey	3	7	26
Peregrine Falcon	2	0	0
Pigeon Hawk	2	1	1
Sparrow Hawk	8	22	89
Unidentified hawks	3,186	2,025	9,290
TOTALS	10,035	10,667	26,646

SOURCE: Compiled from Miller (1952) and Merriam (1953, 1954a, 1954b, 1956).

TABLE 10

CEDAR GROVE AUTUMN HAWK COUNTS (1952–1957)

Species	No. Counted
Turkey Vulture	17
Goshawk	19
Sharp-shinned Hawk	8,524
Cooper's Hawk	268
Red-tailed Hawk	1,407
Red-shouldered Hawk	72
Broad-winged Hawk	15,965
Swainson's Hawk	7
Rough-legged Hawk	39
Golden Eagle	2
Bald Eagle	6
Marsh Hawk	1,115
Osprey	186
Prairie Falcon	1
Peregrine Falcon	150
Pigeon Hawk	798
Sparrow Hawk	370
Unidentified hawks	115
TOTAL	29,061

SOURCE: Compiled from Mueller and Berger (1961).

TABLE 11

DULUTH AUTUMN HAWK COUNTS

Species	1951	1952	1953	1954	1955	1956	1957	1958	1959	1960	1961
Turkey Vulture	12	20	36	36	56	34	26	123	1	84	281
Goshawk	7	24	4	6	16	2	7	5	0	9	69
Sharp-shinned Hawk	686	1,777	973	1,704	3,708	2,043	2,790	5,157	910	2,056	5,997
Cooper's Hawk	136	304	133	49	38	87	31	81	8	38	74
Red-tailed Hawk	71	99	223	45	716	855	291	113	23	636	661
Broad-winged Hawk	3,231	5,338	4,906	1,852	1,638	1,309	11,124	9,991	63	6,864	23,642
Rough-legged Hawk	2	74	2	0	63	14	5	3	1	20	30
Golden Eagle	6	1	1	2	1	0	0	0	0	1	2
Bald Eagle	7	7	0	4	6	6	6	2	0	3	10
Marsh Hawk	69	372	256	57	176	68	240	339	49	117	426
Osprey	20	32	16	26	13	25	39	33	18	39	60
Peregrine Falcon	4	13	5	4	18	4	4	4	2	7	34
Pigeon Hawk	9	50	9	26	39	14	63	50	8	14	20
Sparrow Hawk	53	144	58	366	122	108	245	298	85	95	369
Other hawks	0	3	3	0	0	1	10	5	0	1	2
Unidentified hawks	4,616	5,046	596	335	1,320	569	340	1,898	244	457	965
TOTALS	8,929	13,304	7,221	4,512	7,930	5,139	15,221	18,102	1,412	10,441	32,642
No. Observation Days	4	5	4	6	13	15	12	6	9	17	35

Species	1962	1963	1964	1965	1966	1967	1968	1969	1970	1971	1972
Turkey Vulture	62	26	135	140	19	53	798	209	381	497	282
Goshawk	333	715	31	291	6	7	18	20	7	8	5,152
Sharp-shinned Hawk	3,668	2,006	1,585	3,136	802	2,361	1,874	2,765	6,183	4,106	6,671
Cooper's Hawk	50	37	53	75	13	26	30	32	48	31	96
Red-tailed Hawk	1,223	1,612	844	1,327	279	82	1,072	435	902	1,448	3,621
Broad-winged Hawk	20,603	2,626	10,916	16,257	10,304	7,691	24,768	12,983	62,541	54,410	26,910
Rough-legged Hawk	81	261	25	78	1	2	52	5	67	322	148
Golden Eagle	3	7	2	3	1	0	1	3	7	2	35
Bald Eagle	8	15	9	11	3	2	5	8	40	30	28
Marsh Hawk	126	147	97	102	46	82	222	248	526	515	413
Osprey	45	39	16	31	25	13	37	68	87	90	86
Peregrine Falcon	4	10	3	11	5	1	4	29	8	3	11
Pigeon Hawk	10	7	11	18	3	2	5	19	15	5	13
Sparrow Hawk	217	133	207	258	409	106	227	342	608	542	543
Other Hawks	3	1	0	6	0	0	0	0	1	1	5
Unidentified hawks	279	187	96	512	67	48	63	205	30	100	148
TOTALS	26,715	7,829	14,030	22,256	11,983	10,476	29,176	17,371	71,451	62,110	44,162
No. Observation Days	44	31	18	38	11	11	25	49	66	38	70

SOURCE: Compiled from Hofslund (1966, personal communication) and Sundquist (1973).

TABLE 12

NEW IPSWICH AUTUMN HAWK COUNTS

Species	1952	1953	1954	1955	1956	1957	1958	1960
Sharp-shinned Hawk	233	89	52	16	89	31	. . . .	8
Cooper's Hawk	7	2	5	6	8	2	. . . .	2
Red-tailed Hawk	8	0	3	3	7	0	. . . .	2
Red-shouldered Hawk	0	1	0	0	0	0	. . . .	1
Broad-winged Hawk	949	1,034	1,418	101	1,626	435	361	372
Bald Eagle	1	1	2	1	1	2	. . . .	0
Marsh Hawk	8	2	1	7	14	8	. . . .	3
Osprey	30	22	33	11	48	16	. . . .	10
Sparrow Hawk	52	20	6	8	25	31	. . . .	10
Unidentified hawks	0	1	6	0	0	11	. . . .	4
TOTALS	1,288	1,172	1,526	153	1,818	536	361	412
No. Observation Days	14	6	7	3	10	15	13	5

SOURCE: Compiled from data filed by Cora Wellman with the U.S. Fish and Wildlife Service at Patuxent Wildlife Research Center.

TABLE 13

MOUNT TOM AUTUMN HAWK COUNTS

Species	1952	1953	1955	1957	1958	1959
Turkey Vulture	0	0	3	0	2	1
Sharp-shinned Hawk	99	16	177	144	17	26
Cooper's Hawk	6	8	16	39	5	16
Red-tailed Hawk	4	2	14	17	10	10
Red-shouldered Hawk	112	0	12	14	3	4
Broad-winged Hawk	3,319	271	1,532	1,347	4,229	1,758
Golden Eagle	0	0	1	1	0	1
Bald Eagle	8	1	7	1	4	4
Marsh Hawk	14	0	23	13	8	9
Osprey	65	12	108	72	67	33
Peregrine Falcon	1	0	4	5	0	0
Pigeon Hawk	0	0	3	1	0	3
Sparrow Hawk	31	9	154	109	10	34
Unidentified hawks	18	7	40	0	0	1
TOTALS	3,677	326	2,094	1,763	4,355	1,900

SOURCE: Compiled from data filed by Aaron M. Bagg, Frances T. Elkins, D. Chrisholm Hagar, Mrs. David Riedel, and Robert Smart with the U.S. Fish and Wildlife Service at Patuxent Wildlife Research Center.

TABLE 14

NEW FAIRFIELD AUTUMN HAWK COUNTS

Species	1952	1953	1956	1957	1958	1959	1960	1961
Turkey Vulture	3	0	10	10	4	0	0	6
Sharp-shinned Hawk	40	1	13	33	10	6	11	6
Cooper's Hawk	1	1	1	4	3	0	3	0
Red-tailed Hawk	1	0	2	0	3	2	1	2
Red-shouldered Hawk	3	0	8	6	2	5	4	1
Broad-winged Hawk	1,082	593	3,981	8,255	3,244	1,573	1,951	1,709
Bald Eagle	7	0	9	1	0	2	2	0
Marsh Hawk	8	0	16	1	1	2	3	0
Osprey	11	0	29	25	12	10	30	7
Peregrine Falcon	0	0	0	0	0	1	0	0
Sparrow Hawk	6	0	7	6	5	3	8	6
Unidentified hawks	46	3	123	74	48	78	59	6
TOTALS	1,208	598	4,199	8,415	3,332	1,682	2,072	1,743
No. Observation Days	10	1	12	5	5	6	8	2

SOURCE: Compiled from data filed by Frances J. Gillotti with the U.S. Fish and Wildlife Service at Patuxent Wildlife Research Center.

TABLE 15
WESTPORT AUTUMN HAWK COUNTS

Species	1953	1954	1955	1956	1957	1959	1962
Goshawk	0	0	0	0	0	0	1
Sharp-shinned Hawk	180	100	46	59	89	18	14
Cooper's Hawk	29	40	7	21	14	6	3
Red-tailed Hawk	5	8	4	0	6	0	0
Red-shouldered Hawk	29	10	9	7	17	5	18
Broad-winged Hawk	1,594	1,113	645	336	224	215	2,648
Bald Eagle	2	1	1	1	0	2	2
Marsh Hawk	9	18	2	4	1	6	2
Osprey	36	87	34	23	9	35	35
Peregrine Falcon	2	4	2	0	3	1	0
Pigeon Hawk	10	10	3	5	6	9	2
Sparrow Hawk	39	85	8	31	35	36	31
Unidentified hawks	19	29	2	8	8	7	31
TOTALS	1,954	1,505	763	495	412	340	2,787
No. Observation Days	19	24	10	8	4	18	19

SOURCE: Compiled from data filed by Mortimer F. Brown with the U.S. Fish and Wildlife Service at Patuxent Wildlife Research Center.

TABLE 16

HOOK MOUNTAIN AUTUMN HAWK COUNTS

Species	1966	1967	1968	1969	1970	1971	1972
Turkey Vulture	. . . .	4	20	9	13	. . . .	. . . .
Goshawk	0	0	0	0	4	0	112
Sharp-shinned Hawk	17	50	87	73	481	2,095	1,694
Cooper's Hawk	0	1	0	9	24	37	43
Red-tailed Hawk	2	61	16	31	179	315	452
Red-shouldered Hawk	0	47	0	7	39	76	93
Broad-winged Hawk	254	38	955	149	6,836	3,295	1,105
Rough-legged Hawk	0	0	1	0	4	0	1
Golden Eagle	0	1	0	0	3	4	3
Bald Eagle	1	0	1	1	1	5	1
Marsh Hawk	3	14	27	16	63	200	102
Osprey	4	9	21	13	29	256	122
Peregrine Falcon	0	0	0	0	1	4	7
Pigeon Hawk	0	9	7	0	1	7	7
Sparrow Hawk	6	11	48	38	91	623	324
Unidentified hawks	24	36	314	63	91	102	89
Unidentified eagles	0	0	2	0	1	0	. . . .
TOTALS	311	281	1,499	409	7,861	7,019	4,155
No. Observation Days	3	8	10	10	26	61	74

SOURCE: Compiled from data provided by Mrs. Edward O. Mills, and from Mills and Mills (1971) and Thomas (1971b, 1973).

TABLE 17

JONES BEACH AUTUMN HAWK COUNTS

Species	1958
Sharp-shinned Hawk	21
Cooper's Hawk	1
Marsh Hawk	40
Osprey	37
Peregrine Falcon	12
Pigeon Hawk	78
Sparrow Hawk	405
Unidentified hawks	20
TOTAL	614
No. Observation Days	17

SOURCE: Compiled from Ward (1960b).

TABLE 18

MOUNT PETER AUTUMN HAWK COUNTS

Species	1958	1959	1960	1961	1962	1963	1964	1965	1966	1967	1968	1969	1970	1971	1972
Goshawk	0	0	0	0	0	0	0	0	1	1	0	0	0	0	0
Sharp-shinned Hawk	39	121	74	67	45	71	123	368	164	160	328	305	367	248	469
Cooper's Hawk	6	9	4	10	9	8	13	39	25	6	18	23	26	13	31
Red-tailed Hawk	53	90	75	48	74	19	30	198	213	151	312	244	169	57	202
Red-shouldered Hawk	46	23	16	22	5	1	2	62	44	13	62	27	16	11	9
Broad-winged Hawk	90	1,605	1,293	1,136	3,701	1,428	3,489	1,532	1,393	817	6,658	4,537	3,832	10,944	7,158
Rough-legged Hawk	0	0	0	0	0	0	0	0	0	0	0	0	0	0	0
Golden Eagle	0	0	0	0	0	0	0	0	0	1	0	0	1	0	2
Bald Eagle	1	1	1	1	0	0	4	1	0	0	0	1	4	0	2
Marsh Hawk	19	27	12	24	18	33	17	66	39	33	58	50	42	57	33
Osprey	15	31	25	17	58	32	88	146	61	40	80	99	99	70	89
Peregrine Falcon	2	4	0	2	3	0	0	4	1	4	4	0	1	0	3
Pigeon Hawk	1	6	0	3	1	2	0	3	3	5	4	4	21	1	4
Sparrow Hawk	25	29	61	82	157	201	116	473	538	264	333	322	510	232	168
Unidentified hawks	54	62	97	60	33	74	53	135	48	105	128	105	98	38	66
TOTALS	351	2,008	1,658	1,472	4,104	1,869	3,935	3,027	2,530	1,600	7,986	5,718	5,186	11,669	8,236

Source:: Compiled from Bailey (1967, 1968), Rogers (1971), and Thomas (unpublished data).

TABLE 19

BAKE OVEN KNOB AUTUMN HAWK COUNTS

Species	1957	1961	1962	1963	1964	1965	1966	1967	1968	1969	1970	1971	1972
Turkey Vulture	....	26	47	146	106	113	130	210	182	229	317	241	251
Goshawk	0	3	13	7	9	37	38	24	67	59	79	31	357
Sharp-shinned Hawk	228	71	280	207	429	1,719	1,475	1,954	1,987	2,603	2,517	2,824	2,185
Cooper's Hawk	23	2	18	8	8	58	60	56	65	48	116	65	93
Red-tailed Hawk	135	129	733	881	938	1,276	1,165	1,020	2,149	2,831	3,191	3,572	3,250
Red-shouldered Hawk	3	26	31	44	78	142	325	158	341	226	242	243	186
Broad-winged Hawk	355	236	2,606	5,132	5,940	7,491	6,082	6,278	11,476	7,211	7,729	4,306	8,902
Rough-legged Hawk	0	0	0	5	8	9	2	2	2	5	10	37	5
Golden Eagle	0	2	4	11	13	21	15	17	16	29	34	31	42
Bald Eagle	0	2	3	10	6	25	30	19	21	16	22	23	16
Marsh Hawk	10	15	47	56	100	140	126	136	282	211	369	299	191
Osprey	6	10	110	59	219	379	355	265	297	421	398	444	297
Peregrine Falcon	0	0	5	4	13	10	13	13	6	16	15	17	3
Pigeon Hawk	0	0	3	4	2	13	17	10	8	14	9	7	7
Sparrow Hawk	2	16	69	35	89	199	257	226	241	240	314	239	162
Unidentified hawks	13	23	60	67	85	170	129	173	97	98	141	93	198
Unidentified eagles	0	0	0	0	0	1	0	0	0	0	0	0	0
TOTALS	775	564	4,029	6,676	8,043	11,803	10,219	10,561	17,237	14,257	15,503	12,473*	16,135
No. Observation Days	14	9	20	32	39	54	55	58	56	73	98	100	102

SOURCE: Compiled from Heintzelman (1963a, 1963b, 1966, 1968, 1969b, 1970d), Heintzelman and Armentano (1964), and Heintzelman and MacClay (1972, 1973).

* Includes one Swainson's Hawk.

TABLE 20

BEAR ROCKS AUTUMN HAWK COUNTS

Species	1965	1966	1968	1969
Turkey Vulture	0	0	0	7
Goshawk	15	7	53	11
Sharp-shinned Hawk	143	55	1,163	632
Cooper's Hawk	12	4	30	17
Red-tailed Hawk	172	110	1,037	345
Red-shouldered Hawk	24	10	93	37
Broad-winged Hawk	1,033	3,663	3,075	1,394
Rough-legged Hawk	0	0	2	1
Golden Eagle	3	2	5	2
Bald Eagle	4	6	13	9
Marsh Hawk	9	23	118	31
Osprey	76	112	124	206
Peregrine Falcon	2	0	6	4
Pigeon Hawk	0	0	3	1
Sparrow Hawk	44	40	118	91
Unidentified hawks	8	13	39	14
TOTALS	1,545	4,045	5,879	2,802
No. Observation Days	7	13	29	16

SOURCE: Compiled from data provided by Alan and Paul Grout.

TABLE 21

HAWK MOUNTAIN AUTUMN HAWK COUNTS

Species	1934	1935	1936	1937	1938	1939	1940
Turkey Vulture	166	374	87	44	60	146	150
Goshawk	123	293	177	49	9	26	11
Sharp-shinned Hawk	1,913	4,237	4,486	4,817	3,113	8,529	2,407
Cooper's Hawk	333	553	474	492	204	590	166
Red-tailed Hawk	5,609	4,024	3,177	4,978	2,230	6,496	4,725
Red-shouldered Hawk	90	181	153	163	143	314	149
Broad-winged Hawk	2,026	5,459	7,509	4,500	10,761	5,736	3,159
Rough-legged Hawk	20	9	9	4	....	8	4
Golden Eagle	39	66	54	73	31	83	72
Bald Eagle	52	67	70	38	37	64	38
Marsh Eagle	105	153	149	160	189	273	161
Osprey	31	169	205	201	124	174	91
Peregrine Falcon	25	14	36	41	24	38	25
Pigeon Hawk	19	20	34	10	12	43	11
Sparrow Hawk	13	123	102	141	87	184	60
Unidentified hawks	208	23	11	8	....	....	7
TOTALS	10,772	15,765	16,733	15,719	17,024	22,704	11,236

Species	1941	1942	1946	1947	1948	1949	1950
Turkey Vulture	182	83	64	268	300	376	64
Goshawk	21	9	32	5	14	7	6
Sharp-shinned Hawk	3,909	3,203	2,409	1,745	1,651	2,908	3,719
Cooper's Hawk	416	292	221	122	203	174	263
Red-tailed Hawk	4,700	2,378	2,358	1,677	2,499	2,736	2,669
Red-shouldered Hawk	198	120	248	245	268	283	346
Broad-winged Hawk	5,170	4,362	3,280	7,791	15,454	9,636	6,638
Rough-legged Hawk	2	......	......	......	10	16	2
Golden Eagle	55	35	69	34	40	47	68
Bald Eagle	50	71	42	92	88	98	142
Marsh Hawk	254	107	171	176	186	212	223
Osprey	201	213	191	297	170	212	323
Peregrine Falcon	44	36	26	19	33	33	35
Pigeon Hawk	35	17	20	10	19	20	45
Sparrow Hawk	196	113	98	121	142	334	253
Unidentified hawks	38	38	62	52	96	55	120
TOTALS	15,471	11,077	9,291	12,654	21,173	17,147	14,916

TABLE 21 (Cont.)

HAWK MOUNTAIN AUTUMN HAWK COUNTS

Species	1951	1952	1953	1954	1955	1956	1957	1958	1959	1960
Turkey Vulture	10	19	201	43	62	1	220	14	186	....
Goshawk	21	7	4	96	37	7	81	20	25	15
Sharp-shinned Hawk	3,039	3,559	3,018	3,220	4,381	2,079	2,674	1,765	2,819	2,395
Cooper's Hawk	238	308	165	194	282	122	203	176	170	159
Red-tailed Hawk	2,292	2,555	2,193	2,072	3,802	1,537	2,744	2,956	1,911	2,317
Red-shouldered Hawk	378	347	199	311	428	318	216	470	344	377
Broad-winged Hawk	11,132	12,629	7,270	5,911	9,957	8,748	8,949	8,880	5,282	12,585
Rough-legged Hawk	....	9	2	3	....	10	8	1	12	30
Golden Eagle	54	80	31	40	57	48	46	41	31	38
Bald Eagle	100	114	60	77	89	54	39	46	41	37
Marsh Hawk	222	309	238	151	230	138	187	203	257	290
Osprey	254	343	341	336	359	288	319	321	288	303
Peregrine Falcon	26	32	15	29	35	19	15	25	33	26
Pigeon Hawk	15	45	11	28	23	10	22	19	17	33
Sparrow Hawk	257	210	223	176	276	192	175	219	281	236
Unidentified hawks	67	73	140	93	173	45	98	108	82	52
TOTALS	18,105	20,639	14,111	12,780	20,191	13,616	15,996	15,264	11,779	18,893

Species	1961	1962	1963	1964	1965	1966	1967	1968	1969	1970
Turkey Vulture	129	......	178	7	84	22	3	2	......	......
Goshawk	89	33	28	19	102	30	32	233	146	107
Sharp-shinned Hawk	1,753	2,283	1,528	1,314	3,498	2,947	3,243	3,677	3,708	3,405
Cooper's Hawk	108	77	73	62	103	84	101	220	145	144
Red-tailed Hawk	2,606	2,748	3,474	2,799	3,305	2,578	2,680	4,647	4,338	3,632
Red-shouldered Hawk	333	268	210	223	303	314	374	512	296	350
Broad-winged Hawk	8,403	8,276	9,824	10,218	9,318	10,269	12,052	18,507	12,901	14,372
Rough-legged Hawk	36	7	11	16	4	4	7	6	15	18
Golden Eagle	52	40	28	28	34	16	36	39	28	25
Bald Eagle	51	48	30	28	43	32	38	55	41	28
Marsh Hawk	283	186	178	191	190	178	244	480	321	495
Osprey	352	290	190	328	444	405	457	403	530	600
Peregrine Falcon	23	30	21	21	14	22	22	21	26	27
Pigeon Hawk	13	20	27	11	23	26	20	21	32	23
Sparrow Hawk	470	446	262	237	408	579	666	736	610	521
Unidentified hawks	117	110	101	28	113	311	221	206	282	253
TOTALS	14,818	14,862	16,163	15,530	17,986	17,817	20,196	29,765	23,419	24,000

SOURCE: Compiled from Broun (1949a) and from data in *News Letters to Members* issued by the Hawk Mountain Sanctuary Association.

NOTE: No records were kept during World War II. In 1967 the South Lookout opened, resulting in a considerable increase in the number of hawks counted each autumn thereafter.

TABLE 22

LARKSVILLE MOUNTAIN AUTUMN HAWK COUNTS

Species	1958
Sharp-shinned Hawk	35
Cooper's Hawk	9
Red-tailed Hawk	97
Red-shouldered Hawk	11
Broad-winged Hawk	47
Marsh Hawk	4
Osprey	3
Pigeon Hawk	1
Sparrow Hawk	7
Unidentified hawks	14
TOTAL	228
No. Observation Days	4

SOURCE: Compiled from data filed by Harry Brown, Edwin Johnson, and William Reid with the U.S. Fish and Wildlife Service at Patuxent Wildlife Research Center.

TABLE 23

PIPERSVILLE AUTUMN HAWK COUNTS

Species	1970	1971
Goshawk	1	1
Sharp-shinned Hawk	93	217
Cooper's Hawk	3	0
Red-tailed Hawk	33	33
Red-shouldered Hawk	1	4
Broad-winged Hawk	503	3,751
Rough-legged Hawk	0	2
Golden Eagle	1	0
Marsh Hawk	28	27
Osprey	39	151
Sparrow Hawk	82	243
TOTALS	784	4,429
No. Observation Days	20	28

SOURCE: Compiled from data provided by Ann Webster.

TABLE 24
THE PULPIT AUTUMN HAWK COUNTS

Species	1966	1967	1968	1969	1970	1971	1972
Goshawk	0	0	0	1	2	2	11
Sharp-shinned Hawk	154	46	205	271	166	485	767
Cooper's Hawk	4	3	13	5	6	17	46
Red-tailed Hawk	202	229	351	421	330	643	606
Red-shouldered Hawk	5	13	9	14	5	20	22
Broad-winged Hawk	1,223	77	169	135	551	1,112	1,389
Golden Eagle	0	3	5	2	2	9	12
Bald Eagle	2	0	1	1	0	2	2
Marsh Hawk	22	29	39	18	44	81	69
Osprey	23	14	25	19	34	32	65
Peregrine Falcon	0	1	1	0	0	2	3
Pigeon Hawk	2	0	1	4	0	1	0
Sparrow Hawk	8	10	31	7	21	51	17
Unidentified hawks	12	9	16	16	18	22	55
TOTALS	1,657	434	866	914	1,179	2,479	3,064
No. Observation Days	12	11	17	21	19	38	56

SOURCE: Compiled from data provided by Carl L. Garner.

TABLE 25
ROUTE 183 AUTUMN HAWK COUNTS

Species	1961	1962	1963	1964	1965	1966	1968
Goshawk	0	0	0	0	3	1	2
Sharp-shinned Hawk	46	9	38	6	4	2	5
Cooper's Hawk	1	1	3	1	1	1	1
Red-tailed Hawk	25	51	111	6	29	15	15
Red-shouldered Hawk	3	3	6	3	6	4	1
Broad-winged Hawk	0	95	2	2	0	10	0
Rough-legged Hawk	1	1	0	1	0	0	0
Golden Eagle	0	1	1	0	0	0	0
Marsh Hawk	0	3	0	0	0	0	0
Osprey	0	2	2	1	1	1	0
Peregrine Falcon	0	0	0	0	0	1?	0
Pigeon Hawk	0	0	1	0	0	1	0
Sparrow Hawk	0	2	4	2	1	1	0
TOTALS	76	168	168	22	45	37	24
No. Observation Days	3	6	7	6	4	2	1

SOURCE: Compiled from data provided by Earl L. Poole.

TABLE 26

STERRETT'S GAP AUTUMN HAWK COUNTS

Species	1938	1939	1940	1941
Turkey Vulture	314	333	80	453
Goshawk	21	18	12	5
Sharp-shinned Hawk	2,200	6,489	2,479	3,401
Cooper's Hawk	132	405	130	304
Red-tailed Hawk	2,078	3,263	1,913	3,040
Red-shouldered Hawk	222	226	144	166
Broad-winged Hawk	4,984	2,596	3,150	2,529
Rough-legged Hawk	2	4	1	0
Golden Eagle	28	40	34	42
Bald Eagle	15	24	19	27
Marsh Hawk	100	179	76	226
Osprey	74	89	54	93
Peregrine Falcon	24	43	22	47
Pigeon Hawk	6	21	5	23
Sparrow Hawk	38	53	23	68
Unidentified hawks	17	25	1	8
TOTALS	10,255	13,808	8,143	10,432

1938 = 227 hours of observation, September 9 to November 6.
1939 = 331 hours of observation, September 1 to November 16.
1940 = 257 hours of observation, September 5 to November 16.
1941 = 400 hours of observation, September 5 to November 17.

SOURCE: Compiled from Frey (1943).

TABLE 27

TOTT'S GAP AUTUMN HAWK COUNTS

Species	1958	1959	1960	1961	1962	1965	1972
Goshawk	0	0	0	1	0	0	12
Sharp-shinned Hawk	82	8	32	45	93	2	285
Cooper's Hawk	1	1	1	2	9	0	33
Red-tailed Hawk	5	37	90	25	70	5	401
Red-shouldered Hawk	1	2	0	5	5	0	12
Broad-winged Hawk	0	93	0	91	9	345	875
Golden Eagle	1	1	1	0	2	0	2
Bald Eagle	3	2	0	4	1	3	2
Marsh Hawk	0	3	8	0	1	5	15
Osprey	10	6	6	17	8	37	29
Peregrine Falcon	2	1	1	3	2	0	0
Pigeon Hawk	1	1	0	1	0	0	0
Sparrow Hawk	5	3	0	10	19	7	41
Unidentified hawks	2	3	0	0	0	0	12
TOTALS	113	161	139	204	219	404	1,719
No. Observation Days	3	4	2	5	3	1	27½

SOURCE: Compiled from data provided by Howard Drinkwater, and from Heintzelman (1973).

TABLE 28

WAGGONER'S GAP AUTUMN HAWK COUNTS

Species	1952	1953	1954	1955	1956	1957	1958	1959	1960	1961	1962
Goshawk	0	0	0	0	0	0	0	0	1	0	0
Sharp-shinned Hawk	606	37	88	128	51	62	238	409	255	132	206
Cooper's Hawk	34	6	4	3	10	16	31	38	45	12	22
Red-tailed Hawk	25	4	2	12	7	42	159	124	244	260	201
Red-shouldered Hawk	3	0	2	0	4	2	23	22	38	34	2
Broad-winged Hawk	1,954	116	17	576	230	382	490	944	2,594	515	818
Rough-legged Hawk	0	0	0	0	0	0	0	0	0	1	0
Golden Eagle	2	1	1	2	0	0	8	1	6	3	2
Bald Eagle	5	1	0	10	5	3	20	14	5	18	4
Marsh Hawk	16	7	10	15	5	22	35	32	27	46	16
Osprey	35	2	18	27	9	27	60	59	39	43	54
Peregrine Falcon	7	0	3	0	0	3	3	3	6	6	6
Pigeon Hawk	1	0	0	0	0	0	6	1	2	0	0
Sparrow Hawk	13	2	3	12	2	11	33	29	19	34	30
Unidentified hawks	13	10	4	1	16	0	14	28	110	64	30
TOTALS	2,714	186	152	786	339	570	1,120	1,704	3,391	1,168	1,391
No. Observation Days	6	6	1	3	4	13	24	20	23	27	23

Species	1963	1964	1965	1966	1967	1968	1969	1970	1971	1972
Goshawk	0	0	2	0	0	1	0	2	0	15
Sharp-shinned Hawk	19	51	220	136	200	145	30	234	238	240
Cooper's Hawk	6	10	20	19	18	28	15	26	7	24
Red-tailed Hawk	217	185	85	79	171	193	82	388	148	317
Red-shouldered Hawk	3	4	7	18	5	15	2	45	11	24
Broad-winged Hawk	951	753	690	1,937	1,067	1,722	2,518	2,838	1,062	1,092
Rough-legged Hawk	0	0	0	0	1	1	0	0	0	0
Golden Eagle	1	1	1	0	1	0	2	1	7	7
Bald Eagle	5	6	9	3	4	3	8	3	1	2
Marsh Hawk	14	14	23	11	20	31	14	40	11	36
Osprey	25	25	30	28	28	75	72	71	9	62
Peregrine Falcon	2	1	1	1	1	0	0	2	0	0
Pigeon Hawk	0	1	0	0	2	0	0	0	0	0
Sparrow Hawk	21	25	46	55	54	40	36	61	56	56
Unidentified hawks	23	16	24	28	36	44	53	49	24	74
TOTALS	1,287	1,092	1,158	2,315	1,608	2,298	2,832	3,760	1,574	1,949
No. Observation Days	21	34	19	27	34	23	21	22	19	29

SOURCE: Compiled from data provided by D. L. Knohr, and from data filed by George Salkeld, Evan G. Thomas, Gordon E. Thomas, Marjorie E. Thomas, Charles Throst, and Mrs. Harold Van Riper with the U.S. Fish and Wildlife Service at Patuxent Wildlife Research Center.

TABLE 29

BEARFORT MOUNTAIN AUTUMN HAWK COUNTS

Species	1968	1969	1970
Turkey Vulture	. . .	. . .	15
Goshawk	3	23	8
Sharp-shinned Hawk	26	36	118
Cooper's Hawk	2	16	9
Red-tailed Hawk	93	244	145
Red-shouldered Hawk	31	60	14
Broad-winged Hawk	0	39	2,364
Rough-legged Hawk	0	7	2
Golden Eagle	2	3	2
Bald Eagle	0	3	2
Marsh Hawk	11	15	26
Osprey	2	11	54
Gyrfalcon	0	1	0
Peregrine Falcon	0	0	1
Pigeon Hawk	0	4	2
Sparrow Hawk	7	15	244
Unidentified hawks	6	48	40
Unidentified eagles	0	1	0
TOTALS	183	526	3,046
No. Observation Days	5	28	40

SOURCE: Compiled from Koebel (1970).

TABLE 30

HAWKS SHOT AT CAPE MAY POINT (AUTUMN 1935)

Species	Number Passing	Number Shot
Turkey Vulture	1,678	. . . .
Sharp-shinned Hawk	8,206	1,008
Cooper's Hawk	840	62
Red-tailed Hawk	50	0
Red-shouldered Hawk	12	0
Broad-winged Hawk	367	3
Rough-legged Hawk	2	0
Bald Eagle	60	0
Marsh Hawk	274	1
Osprey	706	1
Peregrine Falcon	56	1
Pigeon Hawk	402	3
Sparrow Hawk	777	1
Unidentified hawks	22	0
TOTALS	13,452	1,080

SOURCE: Compiled from Stone (1937).

TABLE 31

CAPE MAY POINT AUTUMN HAWK COUNTS

Species	1931	1932	1935	1965	1970	1971	1972
Turkey Vulture	.....	.....	1,678	152	649	344	303
Black Vulture	.....	.....	.....	....	5	0	.....
Goshawk	.....	.....	.....	2	5	0	26
Sharp-shinned Hawk	10,000	5,765	8,206	909	5,557	6,115	7,910
Cooper's Hawk	500	1,222	840	53	445	152	178
Red-tailed Hawk	.....	177	50	374	943	1,046	997
Red-shouldered Hawk	.....	600	12	117	126	62	29
Broad-winged Hawk	2,000	400	367	87	1,863	567	359
Rough-legged Hawk	.....	.....	2	2	5	10	2
Golden Eagle	0	0	0	0	6	1	4
Bald Eagle	40	10	60	6	5	3	1
Marsh Hawk	100	264	274	192	868	308	279
Osprey	common	102	706	90	305	219	181
Peregrine Falcon	20	42	56	19	111*	50	44
Pigeon Hawk	1,200	1,707	402	66	234	407	525
Sparrow Hawk	200	322	777	1,725	30,290	7,132	4,838
Unidentified hawks	.....	.....	22	157	156	0	0
TOTALS	14,060	10,611	13,452	3,951	41,573	16,416	15,676
No. Observation Days	.....	.....	.....	.....	.....	117	86

SOURCE: Compiled from Allen and Peterson (1936), Choate (*in* Heintzelman, 1970e), Choate (1972), Choate and Tilly (1973), and Clark (1972, 1973).

* I question the accuracy of this count; Clark (1971) reported a total of 40 Peregrine Falcons based upon partial coverage.

TABLE 32

CATFISH FIRE TOWER AUTUMN HAWK COUNTS

Species	1970	1971
Turkey Vulture	17	82
Goshawk	9	0
Sharp-shinned Hawk	84	140
Cooper's Hawk	5	2
Red-tailed Hawk	51	104
Broad-winged Hawk	2,871	525
Golden Eagle	2	1
Marsh Hawk	19	12
Osprey	28	44
Peregrine Falcon	0	2
Pigeon Hawk	2	4
Sparrow Hawk	20	16
Unidentified hawks	8	8
TOTALS	3,116	940
No. Observation Days	5	14

Source: Compiled from Heintzelman (1972a, 1972b).

TABLE 33

HIGH MOUNTAIN AUTUMN HAWK COUNTS

Species	1965	1966	1967
Sharp-shinned Hawk	9	11	1
Cooper's Hawk	1	1	0
Red-tailed Hawk	0	1	0
Red-shouldered Hawk	1	0	0
Broad-winged Hawk	197	439	59
Bald Eagle	0	0	2
Marsh Hawk	1	1	1
Osprey	7	16	3
Peregrine Falcon	0	2	0
Sparrow Hawk	14	9	15
Unidentified hawk	2	1	3
TOTALS	232	481	84
No. Observation Days	2	5	5

Source: Compiled from data provided by Irving H. Black.

TABLE 34

MONTCLAIR SANCTUARY AUTUMN HAWK COUNTS

Species	1957	1958	1959	1960	1961	1962	1963	1964	1965	1966	1967	1968	1969	1970	1971	1972
Sharp-shinned Hawk	462	216	107	231	153	142	157	194	221	86	117	246	136	687	820	826
Cooper's Hawk	26	22	14	14	8	9	12	9	9	8	8	13	6	23	19	17
Red-tailed Hawk	25	16	13	22	13	35	6	9	23	12	22	45	16	43	48	87
Red-shouldered Hawk	13	6	5	2	15	20	11	11	7	9	2	6	11	7	12	7
Broad-winged Hawk	517	6,167	2,346	1,321	2,415	11,492	1,024	990	7,768	1,606	1,785	6,917	1,601	11,458	3,814	1,734
Rough-legged Hawk	0	1	0	0	0	0	0	0	0	0	0	0	0	0	0	1
Golden Eagle	0	0	0	0	0	0	0	2	0	1	0	0	0	0	0	0
Bald Eagle	1	7	7	2	4	1	1	2	7	0	4	3	2	5	2	4
Marsh Hawk	35	34	36	58	28	39	27	46	45	23	36	34	31	90	45	35
Osprey	161	189	142	136	122	296	70	117	133	157	103	95	157	282	280	210
Peregrine Falcon	8	9	15	6	6	6	11	7	0	0	3	0	2	1	4	1
Pigeon Hawk	9	9	17	6	4	8	10	6	4	9	4	5	7	5	6	6
Sparrow Hawk	357	223	221	397	342	275	363	246	588	366	399	671	437	729	1,008	608
Unidentified hawks	146	59	105	85	54	78	43	44	44	33	26	26	23	68	42	48
TOTALS	1,760	6,958	3,028	2,280	3,164	12,401	1,735	1,681	8,849	2,310	2,509	8,061	2,429	13,398	6,100	3,584
No. Observation Days	20	27	26	26	23	23	26	23	29	22	23	24	25	38	41	46

SOURCE: Compiled from Redmond and Breck (1961), Breck (1962, 1963), Breck and Breck (1964, 1965, 1966), and Bihun (1967, 1968, 1969, 1970, 1972, 1973).

RACCOON RIDGE AUTUMN HAWK COUNTS

Species	1939	1940	1941	1951	1952	1954	1959	1968	1969	1970	1971	1972
Turkey Vulture	5	2	0	0	0	0	0	0	0	39	....	....
Goshawk	4	7	7	0	0	0	1	2	0	17	33	202
Sharp-shinned Hawk	247	89	60	3	118	32	1	12	0	236	789	1,490
Cooper's Hawk	18	2	1	0	6	1	1	1	0	10	57	46
Red-tailed Hawk	577	469	19	4	31	2	70	23	0	911	3,107	2,526
Red-shouldered Hawk	90	21	11	0	14	0	0	2	0	37	67	105
Broad-winged Hawk	210	0	0	0	0	736	0	565	0	21	1,802	10,071
Rough-legged Hawk	2	1	0	0	0	0	1	0	0	3	5	12
Golden Eagle	3	2	1	0	0	0	2	1	0	7	19	26
Bald Eagle	19	0	0	0	0	1	0	0	1	4	4	8
Marsh Hawk	11	10	1	3	8	3	12	9	0	29	60	134
Osprey	16	2	5	0	1	7	0	44	5	22	85	289
Peregrine Falcon	1	1	1	0	1	0	0	0	0	1	6	9
Pigeon Hawk	0	0	5	0	0	0	0	0	0	1	5	12
Sparrow Hawk	17	1	3	0	3	9	0	2	3	3	147	329
Unidentified hawks	0	3	0	0	13	0	0	0	0	30	37	8
TOTALS	1,220	610	114	10	195	791	88	661	9	1,371	6,223	15,267
No. Observation Days	6	2	3	1	3	1	1	3	1	15	43	75

SOURCE: Compiled from data provided by William Rusling, Howard Drinkwater, Clarence D. Brown, and Fred Tilly (1972a, 1972b).

TABLE 36

SUNRISE MOUNTAIN AUTUMN HAWK COUNTS

Species	1953	1954	1955	1956	1957	1958	1960	1965	1967	1968	1969
Sharp-shinned Hawk	29	12	14	41	8	17	11	5	25	29	0
Cooper's Hawk	0	1	0	7	1	3	1	0	0	1	0
Red-tailed Hawk	2	14	21	15	7	9	3	22	10	7	0
Red-shouldered Hawk	0	4	10	1	0	7	0	7	1	2	0
Broad-winged Hawk	38	0	0	453	182	208	253	0	25	71	10
Bald Eagle	3	0	0	2	1	0	0	0	0	0	0
Golden Eagle	0	0	0	0	0	0	1	0	0	0	0
Marsh Hawk	5	0	7	1	1	7	0	1	1	9	0
Osprey	16	0	2	16	3	9	10	0	4	7	1
Peregrine Falcon	1	0	0	5	1	0	0	0	0	0	0
Pigeon Hawk	0	0	0	0	0	0	0	0	0	1	0
Sparrow Hawk	0	0	0	11	3	1	13	4	10	46	10
Unidentified hawks	0	0	0	0	1	7	0	0	3	0	0
TOTALS	94	31	54	552	208	268	292	39	79	173	21

SOURCE: Compiled from data provided by Howard Drinkwater and Fred Tilly, and from data filed by Allen and Ella Black with the U.S. Fish and Wildlife Service at Patuxent Wildlife Research Center.

TABLE 37

MARYLAND AUTUMN HAWK COUNTS (17 SEPTEMBER 1949)

Species	Locations							
	A	B	C	D	E	F	G	H
Sharp-shinned Hawk	....	....	2	7	0	8	1	0
Cooper's Hawk	....	....	1	0	0	1	5	0
Red-tailed Hawk	....	....	2	19	0	12	4	4
Red-shouldered Hawk	....	....	0	16	0	1	0	1
Broad-winged Hawk	....	....	687	350	322	263	310	30
Bald Eagle	....	....	0	2	0	0	0	1
Marsh Hawk	....	....	0	0	0	1	0	0
Osprey	....	....	0	0	0	1	0	1
Sparrow Hawk	....	....	1	2	0	4	0	0
Unidentified hawks (*Buteo* sp.)	....	....	0	109	4	14	5	217
TOTALS	250*	1,216*	693	505	326	305	325	254

A = Dans Mountain   E = Sideling Mountain
B = Wills Mountain   F = Cross Mountain
C = Nicholas Mountain   G = Foxville Tower
D = Martin Mountain   H = Monument Knob

SOURCE: Compiled from Robbins (1950).
* Quantitative records for each species are unavailable, but most of the birds were Broad-winged Hawks.

TABLE 38

ASSATEAGUE ISLAND AUTUMN PEREGRINE FALCON COUNTS
(BIRDS TRAPPED)

Year	No. Peregrines Recorded	References
1959	47	Berry (1971)
1960	23	Berry (1971)
1961	27	Berry (1971)
1962	35	Berry (1971)
1963	43	Berry (1971)
1964	33	Berry (1971)
1965	30	Berry (1971)
1966	45*	Berry (1971)
1967	63*	Berry (1971)
1968	30*	Berry (1971)
1969	34*	Berry (1971)
1970	68	Ward and Berry (1972)
1971	120	Ward and Berry (1972)

SOURCE: Compiled from Berry (1971) and Ward and Berry (1972).
* J. L. Ruos (unpublished data) presents the following numbers of Peregrine Falcons observed on Assateague Island: 1966 = 46, 1967 = 17, 1968 = 29, and 1969 = 19. Berry (1971: 42) states that 143 birds actually may have been sighted in 1969, but that this figure is probably inflated by 20 percent or more.

TABLE 39

COVE POINT AUTUMN HAWK COUNTS

Species	1949
Sharp-shinned Hawk	111
Cooper's Hawk	10
Red-tailed Hawk	3
Broad-winged Hawk	2,355
Bald Eagle	7
Marsh Hawk	14
Osprey	3
Peregrine Falcon	1
Pigeon Hawk	3
Sparrow Hawk	31
Unidentified hawks	8
TOTAL	2,546
No. Observation Days	10

SOURCE: Compiled from Robbins (1950).

TABLE 40

MONUMENT KNOB AUTUMN HAWK COUNTS

Species	1949	1950	1964	1965	1966	1967	1968
Goshawk	0	2	0	0	0	0	0
Sharp-shinned Hawk	168	263	26	6	38	23	7
Cooper's Hawk	14	15	0	0	1	0	0
Red-tailed Hawk	339	455	7	0	11	36	0
Red-shouldered Hawk	52	65	0	0	4	3	0
Broad-winged Hawk	414	1,986	24	10	139	15	5
Rough-legged Hawk	2	0	0	0	0	0	0
Golden Eagle	8	7	1	0	0	0	0
Bald Eagle	2	3	0	0	0	0	0
Marsh Hawk	43	25	4	0	7	0	0
Osprey	11	19	5	0	12	4	1
Peregrine Falcon	3	10	0	0	0	0	0
Pigeon Hawk	0	0	0	0	0	1	0
Sparrow Hawk	11	33	2	0	6	0	0
Unidentified hawks	217	6	0	0	0	0	0
TOTALS	1,284	2,889	69	16	218	82	13
No. Observation Days	13	15	3	1	4	3	1

SOURCE: Compiled from Beaton (1951) and from data provided by Herbert E. Douglas.

TABLE 41

SHENANDOAH MOUNTAIN AUTUMN
HAWK COUNTS (1949–1959)

Species	Number Counted
Sharp-shinned Hawk	21
Cooper's Hawk	23
Red-tailed Hawk	25
Red-shouldered Hawk	156
Broad-winged Hawk	2,015
Rough-legged Hawk	6*
Bald Eagle	8
Marsh Hawk	5
Osprey	24
Peregrine Falcon	1
Sparrow Hawk	2
Unidentified hawks	208
TOTAL	2,494

SOURCE: Compiled from Carpenter (1960).
* That Rough-legged Hawks occurred during September hawk flights in Virginia is seriously doubted.

TABLE 42

CAPE CHARLES AUTUMN HAWK
COUNT (1936)

Species	Number Counted
Sharp-shinned Hawk	6,932
Cooper's Hawk	2,608
Red-tailed Hawk	297
Red-shouldered Hawk	171
Broad-winged Hawk	570
Rough-legged Hawk	2
Bald Eagle	106
Marsh Hawk	350
Osprey	600
Peregrine Falcon	242
Pigeon Hawk	416
Sparrow Hawk	913
TOTAL	13,207

SOURCE: Compiled from Rusling (1936).

TABLE 43

HANGING ROCKS FIRE TOWER AUTUMN HAWK COUNTS

Species	1952	1955	1956	1957	1958	1959	1960	1961	1962	1963	1964	1965	1966	1967	1968	1969
Sharp-shinned Hawk	2	1	13	14	10	1	4	15	17	24	20	30	5	2	6	5
Cooper's Hawk	3	0	5	18	6	3	3	10	15	14	1	19	14	3	4	4
Red-tailed Hawk	1	0	4	6	3	5	7	24	4	36	20	12	6	1	5	10
Red-shouldered Hawk	1	0	4	9	2	5	1	4	3	7	2	8	25	3	1	4
Broad-winged Hawk	1,219	21	586	3,636	270	527	745	325	3,280	60	495	239	1,029	854	1,488	440
Golden Eagle	0	0	0	2	0	0	0	0	0	2	0	0	0	0	0	0
Bald Eagle	0	0	0	1	0	0	0	1	0	0	0	0	0	0	0	0
Marsh Hawk	0	1	0	0	0	0	1	9	5	0	3	5	4	2	4	4
Osprey	2	2	2	3	9	0	1	2	6	3	11	2	3	4	5	14
Pigeon Hawk	0	0	0	0	0	0	0	0	0	0	0	0	1	0	0	0
Sparrow Hawk	8	1	4	30	6	0	4	11	5	9	10	14	19	9	4	5
Unidentified hawks	14	8	28	0	11	2	4	55	6	15	11	12	8	3	7	2
TOTALS	1,250	34	646	3,719	317	543	770	456	3,341	172	573	341	1,114	881	1,524	488
No. Observation Days	1	1	2	2	2	2	3	2	3	6	8	6	6	4	4	7

SOURCE: Compiled from Hurley (1970).

TABLE 44

TENNESSEE BROAD-WINGED HAWK COUNTS (1951–1971)

Year	Number of Hawks Counted
1951	3,911
1952	2,464
1953	896
1954	1,920
1955	532
1956	981
1957	2,885
1958	1,564
1959	10,998
1960	10,135
1961	5,915
1962	5,915
1963	5,434
1964	2,954
1965	15,583
1966	7,255
1967	4,635
1968	24,036
1969	4,215
1970	16,598
1971	13,452
TOTAL	142,278

SOURCE: Compiled from Behrend (1951–1954b), Finucane (1956–1972), and Odom (1966).

TABLE 45

ASPECT-RATIOS OF RAPTOR WINGS

Species	Aspect-Ratio	Number of Specimens
Turkey Vulture	2.65±0.04	..
Sharp-shinned Hawk	2.18	..
Cooper's Hawk	2.19	..
Red-shouldered Hawk	2.24±0.06	8
Broad-winged Hawk	2.28±0.02	7
Osprey	3.00	..
Peregrine Falcon	3.06	..
Sparrow Hawk	2.63±0.04	11

SOURCE: Compiled from Hartman (1961).

TABLE 46

RAPTOR WING LOADING

Species	Sex	Number of Specimens	Weight (Grams)	Wing Area (Sq. Cm.)	Wing Loading (Gms./Sq. Cm.)	Reference
Turkey Vulture	?	1	2,409	4,356	0.54	Poole (1938)
Goshawk	M	1	818	1,462	0.56	Brown and Amadon (1968)
Goshawk	M	1	848.6	1,480	0.57	Poole (1938)
Goshawk	F	1	1,136	1,976	0.58	Brown and Amadon (1968)
Goshawk	F	1	1,370	2,004	0.68	Poole (1938)
Sharp-shinned Hawk	M	9	99	412	0.24	Brown and Amadon (1968)
Sharp-shinned Hawk	M	2	97.5	439	0.22	Poole (1938)
Sharp-shinned Hawk	F	12	171	560	0.31	Brown and Amadon (1968)
Sharp-shinned Hawk	F	1	171	607	0.28	Poole (1938)
Cooper's Hawk	M	3	295	804	0.37	Brown and Amadon (1968)
Cooper's Hawk	F	10	441	1,064	0.41	Brown and Amadon (1968)
Cooper's Hawk	?	2	428.5	898	0.47	Poole (1938)
Red-tailed Hawk	M	1	875	1,878	0.46	Poole (1938)
Red-tailed Hawk	F	2	1,307	2,294	0.56	Poole (1938)
Red-shouldered Hawk	?	3	804	1,656	0.48	Poole (1938)
Broad-winged Hawk	?	3	376	1,012	0.37	Poole (1938)
Rough-legged Hawk	?	8	942	2,342	0.40	Brown and Amadon (1968)
Rough-legged Hawk	?	1	1,110	2,592	0.42	Poole (1938)
Golden Eagle	F	1	4,664	6,520	0.70	Poole (1938)
Golden Eagle	?	1	3,712	5,382	0.67	Brown and Amadon (1968)
Bald Eagle						
Marsh Hawk	M	1	331	1,397	0.24	Brown and Amadon (1968)
Marsh Hawk	M	1	414	1,382	0.29	Poole (1938)
Marsh Hawk	F	1	472	1,761	0.26	Brown and Amadon (1968)
Marsh Hawk	F	1	615	1,696	0.30	Poole (1938)
Osprey	M	3	1,350	2,633	0.51	Brown and Amadon (1968)
Osprey	F	2	2,001	3,082	0.65	Brown and Amadon (1968)
Osprey	?	2	1,797.5	3,211	0.55	Poole (1938)
Peregrine Falcon	M	1	712	1,146	0.62	Poole (1938)
Peregrine Falcon	F	1	1,222.5	1,342	0.91	Poole (1938)
Peregrine Falcon	?	1	813	1,301	0.63	Brown and Amadon (1968)
Pigeon Hawk	M	1	173	410	0.42	Poole (1938)
Pigeon Hawk	?	1	145	435	0.33	Brown and Amadon (1968)
Sparrow Hawk	F	2	137	372	0.36	Poole (1938)

SOURCE: Compiled from Brown and Amadon (1968) and Poole (1938).

NOTE: Poole's data have been reworked for use here.

TABLE 47

WEIGHT AND WING AREA OF RAPTOR WINGS

Species	Sex	Number of Specimens	Weight (Grams)	Wing Area (Sq. Cm.)	Wing Weight (Sq. Cm./ Grams)
Turkey Vulture	?	1	2,409.0	4,356.0	1.81
Goshawk	M	1	848.6	1,480.0	1.74
Goshawk	F	1	1,370.0	2,004.0	1.45
Sharp-shinned Hawk	M	2	97.5	439.0	4.50
Sharp-shinned Hawk	F	1	171.0	609.0	3.55
Cooper's Hawk	?	2	428.5	898.0	2.07
Red-tailed Hawk	M	1	875.0	1,878.0	2.14
Red-tailed Hawk	F	2	1,307.0	2,294.0	1.75
Red-shouldered Hawk	?	3	804.0	1,656.0	2.11
Broad-winged Hawk	?	1	376.0	1,012.0	2.69
Rough-legged Hawk	?	1	1,110.0	2,592.0	2.33
Golden Eagle	F	1	4,664.0	6,520.0	1.39
Marsh Hawk	M	1	414.0	1,382.0	3.34
Marsh Hawk	F	1	615.0	1,696.0	2.75
Osprey	?	2	1,797.5	3,211.0	1.79
Peregrine Falcon	M	1	712.0	1,146.0	1.61
Peregrine Falcon	F	1	1,222.5	1,342.0	1.10
Pigeon Hawk	M	1	173.0	410.0	2.37
Sparrow Hawk	F	2	137.0	372.0	2.74

SOURCE: Compiled from Poole (1938).

TABLE 48

TAIL AREAS, GLIDE AREAS, AND BUOYANCY INDICES

Species	Tail Area (Cm.²/Gram)	Glide Area (Cm.²/Gram)	Buoyancy Index
Turkey Vulture	0.43±0.01	3.90±0.16	5.81
Sharp-shinned Hawk	1.14	4.84	4.40
Cooper's Hawk	0.54	3.32	4.50
Red-shouldered Hawk	0.56±0.05	3.19±0.23	4.76
		4.00±0.22	
Broad-winged Hawk	0.59±0.07	3.59±0.17	4.35
Osprey	0.18	2.49	5.05
Peregrine Falcon	0.37	2.19	3.98
Sparrow Hawk	1.05±0.06	4.92±0.16	4.04

SOURCE: Compiled from Hartman (1961).

TABLE 49

COMPARATIVE HEART WEIGHTS

Species	Sex	Number of Specimens	Body Weight (Grams)	Number of Specimens	Heart Weight (% of Body Weight)
Turkey Vulture	M	15	1426±10	15	0.74±0.02
Turkey Vulture	F	5	1589±118.35	. .	. . . .
Black Vulture	M	5	2065±76	. .	. . . .
Black Vulture	F	1	1950	8	0.89±0.03
Sharp-shinned Hawk	F	1	171.0		0.73
Cooper's Hawk	M	1	315		0.81
Cooper's Hawk	F	6	535±19		. . . .
Red-shouldered Hawk	M	10	475±25.6	7	0.55±0.03
Red-shouldered Hawk	F	14	643±25.7		. . . .
Broad-winged Hawk	M	12	359.6±8.7	9	0.54±0.02
Broad-winged Hawk	F	4	412±14		. . . .
Osprey	M	2	1530, 1500	1	0.84
Osprey	F	3	1837±235		. . . .
Peregrine Falcon	F	2	825, 825	1	1.23
Sparrow Hawk	M	4	85.7±7.42	15	1.01±0.03
Sparrow Hawk	F	11	112.5±3.43		

SOURCE: Compiled from Hartman (1961).

TABLE 50

STERNUM AND RELATED MEASUREMENTS

Measurements	A. s. velox		A. cooperii		A. g. atricapillus	
	Male	Female	Male	Female	Male	Female
Total Length of Wing (Mm)	(9) 140.5	(12) 167.9	(9) 204.3	(10) 232.9	(7) 297.1	(8) 315.0
Wing Area (Square Centimeters)	(5) 411.6	(8) 559.8	(3) 803.6	(10) 1064.2	(1) 1462.2	(1) 1976.4
Weight (Grams)	(42) 98.8	(37) 171.4	(24) 294.9	(29) 440.6	(14) 818.4	(21) 1136.8
Sternum "length" x width x depth	(11) 3.05	(13) 5.28	(15) 11.44	(12) 16.81	(10) 36.55	(11) 41.15
Wing Loading (Gm/Sq. Centimeter)	2.40	3.06	3.67	4.14	5.60	5.75

SOURCE: Compiled from R. W. Storer (1955) and Poole (1938).

NOTE: The wing loading values have been printed as they appear in the source. The decimal point should be moved one place to the left. The size of each sample is indicated within parentheses.

TABLE 51

HAWK MUSCLE WEIGHTS

Species	Pectoral & Supracoracoideus Percent Body Weight		Pectoralis Percent Body Weight		Supracoracoideus Percent Body Weight	
Turkey Vulture (Florida)	(11)	15.8±0.21 P<0.01		....		....
Turkey Vulture (Panama)	(4)	17.45±0.51	(3)	16.4±0.64	(3)	1.26±0.03
Sharp-shinned Hawk		22.2		....		....
Cooper's Hawk		13.3		....		....
Red-shouldered Hawk	(3F)	13.60±0.88		....		....
	(5M)	11.31±0.74		....	(1)	0.47
Broad-winged Hawk	(7)	13.76±0.57	(5)	12.81±0.58	(5)	0.42±0.05
Osprey		14.69		14.0		0.69
Peregrine Falcon		19.20		18.2		0.82
Sparrow Hawk	(13)	15.0±0.33	(7)	14.82±0.39	(7)	0.59±0.03

SOURCE: Compiled from Hartman (1961).
NOTE: The size of each sample is indicated within parentheses.

TABLE 52

FLIGHT SPEEDS OF HAWKS MIGRATING PAST HAWK MOUNTAIN

Species	Number	Range of Speeds (Miles per Hour)	Mean Speed (Miles per Hour)
Turkey Vulture	1	34	34
Goshawk	1	38	38
Sharp-shinned Hawk	37	16–60	30.0
Cooper's Hawk	12	21–55	29.3
Red-tailed Hawk	54	20–40	29.0
Red-shouldered Hawk	7	18–34	28.3
Broad-winged Hawk	8	20–40	31.7
Golden Eagle	2	28–32	30.0
Bald Eagle	2	36–44	40.0
Marsh Hawk	4	21–38	28.7
Osprey	16	20–80	41.5
Peregrine Falcon	3	28–32	30.0
Pigeon Hawk	1	28	28
Sparrow Hawk	4	22–36	26.2

SOURCE: Compiled from Broun and Goodwin (1943).

TABLE 53

DAILY RHYTHM OF MIGRATING HAWKS
BAKE OVEN KNOB, PENNSYLVANIA (1968–1971)

Species	Number	0700	0800	0900	1000	1100	1200	1300	1400	1500	1600	1700
Accipiters	10,170	0.5	5	13	18	16	15	15	11	5	1	0.1
Buteos	12,746	....	1	6	9	13	14	16	20	16	5	..
Falcons	1,110	....	6	11	15	12	14	14	13	11	4	..
Marsh Hawk	1,138	1	4	10	15	15	14	13	13	10	4	1
Osprey	1,546	1	3	6	8	11	16	13	17	14	9	2

SOURCE: Compiled from Heintzelman's data.

NOTE: The figure provided for each hour interval (e.g., 0800–0900) represents the percentage of the total observed (No.).

TABLE 54

SEPTEMBER SHARP-SHINNED HAWK FLIGHT RHYTHMS AT
BAKE OVEN KNOB

Year	No. of Days	Total No. Hawks	Morning Flights		Afternoon Flights	
			No. Hawks	% Total Flight	No. Hawks	% Total Flight
1964	17	170	77	45	93	55
1965	16	201	80	40	121	60
1966	16	126	68	54	58	46
1967	20	240	122	51	118	49
1968	18	418	239	57	179	43
1969	21	215	113	53	102	47
1970	25	686	334	49	352	51
1971	26	346	138	40	208	60
TOTALS	159	2,402	1,171	49	1,231	51

Source: Compiled from Heintzelman's data.

TABLE 55

DAILY RHYTHM OF MIGRATING EAGLES
BAKE OVEN KNOB, PENNSYLVANIA (1961–1972)

Species	Number	0700	0800	0900	1000	1100	1200	1300	1400	1500	1600	1700
Golden Eagle	234	.....	1	8	14	9	13	20	15	14	5	1
Bald Eagle	192	.....	1	6	11	20	18	14	14	9	6	1

SOURCE: Compiled from Heintzelman's data.

NOTE: The figure provided for each hour interval (e.g., 0800–0900) represents the percentage of the total observed (No.).

TABLE 56

SHARP-SHINNED HAWK FLIGHTS
KITTATINNY RIDGE (1972)

Date	Raccoon Ridge	Bake Oven Knob	Hawk Mountain
Sept. 25	. . .	200	102
Sept. 26	. . .	95	103
Sept. 27	58	. . .	140
Sept. 28	19	. . .	121
Oct. 1	41	63	119
Oct. 2	. . .	80	118
Oct. 4	. . .	23	117
Oct. 8	64	183	224
Oct. 9	109	152	234
Oct. 11	185	215	272
Oct. 14	59	246	185
Oct. 15	172	168	392
Oct. 17	165	61	449
Oct. 24	39	56	100

SOURCE: Compiled from Heintzelman and MacClay (1973), Sharadin (1973), and Tilly (1973).

TABLE 57

COMPOSITE BALD EAGLE COUNT
KITTATINNY RIDGE

Year	No. Eagles Counted		Composite Eagle Count — Bake Oven Knob and Hawk Mountain	Bake Oven Knob Eagles Not Seen at Hawk Mountain
	Bake Oven Knob	Hawk Mountain		
1961	2	43	44	50.0%
1962	3	33	35	66.6
1963	10	24	31	70.0
1964	6	29	30	16.7
1965	25	36	48	48.0
1966	30	31	52	70.0
1967	19	38	45	36.8

SOURCE: Compiled from Hawk Mountain Sanctuary Association data, and from Heintzelman data.

TABLE 58

AGE RATIOS OF
GOLDEN EAGLES PASSING BAKE OVEN KNOB

Year	Number of Adults	Percent	Number of Subadults or Immatures	Percent
1961	1	50.0	1	50.0
1962	2	50.0	2	50.0
1963	7	63.5	4	36.5
1964	7	53.8	6	46.2
1965	15	71.5	6	28.5
1966	12	80.0	3	20.0
1967	12	70.6	5	29.4
1968	8	50.0	8	50.0
1969	21	72.5	8	27.5
1970	24	70.6	10	29.4
1971	13	41.9	18	58.1
TOTALS	122	63.2	71	36.8

SOURCE: Compiled from Heintzelman and MacClay (1972).

TABLE 59

AGE RATIOS OF
BALD EAGLES PASSING BAKE OVEN KNOB

Year	Number of Adults	Percent	Number of Immatures	Percent
1961	2	100.0	0	0.0
1962	2	66.7	1	33.3
1963	9	90.0	1	10.0
1964	5	83.3	1	16.7
1965	18	72.0	7	28.0
1966	25	83.3	5	16.7
1967	17	89.4	2	10.6
1968	17	80.9	4	19.1
1969	14	87.5	2	12.5
1970	16	72.7	6	27.3
1971	19	82.6	4	17.4
TOTALS	144	81.4	33	18.6

SOURCE: Compiled from Heintzelman and MacClay (1972).

TABLE 60

MEASUREMENTS OF ADULT BROAD-WINGED HAWKS

Subspecies	Sex	Number	Wing	Tail	Culmen from Cere	Tarsus
B. p. platypterus	M	17	244-277 (262.8)	148-173.5 (159)	17-20 (18.2)	57.5-65.5 (62.3)
B. p. platypterus	F	17	265-296 (282.8)	155-185.4 (171.2)	17.1-20.5 (19.3)	59-66.4 (62.8)
B. p. cubanensis	M	2	250-256	150-156.6	19.5-19.7	65
B. p. cubanensis	F	3	254-266 (261)	160.1-164 (161.3)	19.5-21.3 (20.2)	59.5-61.5 (60.7)
B. p. brunnescens	F	1	264.5	159	20.8	56.3
B. p. insulicola	M	1	227	145.9	18.5	56
B. p. rivierei	M	1	245	149	19.3	55.5
B. p. rivierei	F	1	250	145	18	54.5
B. p. antillarum	M	4	254.5-268 (260.4)	148.1-163.5 (154.9)	18.4-19.8 (18.8)	53.2-57.9 (55.1)
B. p. antillarum	F	3	260.8-272.6 (267.4)	153.6-166.6 (158.9)	19.3-20.5 (19.9)	54-57 (55.4)

SOURCE: Compiled from Friedmann (1950).

NOTE: All measurements are in millimeters. The numbers within parentheses represent the average length of measurements provided in the complete original data.

Literature Cited

Allen, R. P., and R. T. Peterson
   1936   The Hawk Migrations at Cape May Point, New Jersey. *Auk*, 53: 393–404.
Amadon, D.
   1967   Birds of Prey in the Collection of the American Museum of Natural History. *Raptor Research News*, 1 (3): 53–54.
American Ornithologists' Union
   1957   Check-list of North American Birds. Fifth Edition. Port City Press, Inc., Baltimore, Md.
Anonymous
   1953   Hawk Count—1952. *New Hampshire Bird News*, 6 (1): 19.
   1972   No title; summary of some northern hawk watching sites. News Letter to Members No. 44. Hawk Mountain Sanctuary Association, Kempton, Pa. Pp. 36–37.
Austin, O. L., Jr.
   1932   The Birds of Newfoundland Labrador. Memoirs no. 7. Nuttall Ornithological Club, Cambridge, Mass.
Austing, G. R.
   1964   The World of the Red-tailed Hawk. J. B. Lippincott Co., Philadelphia, Pa.

Bagg, A. M.
1947 Watch New England Hawk Flights! *Bulletin Massachusetts Audubon Society,* 31: 81–82.
1949 Flight Over the Valley. Connecticut Valley Hawk Flights— September, 1948. *Bulletin Massachusetts Audubon Society,* 33: 135–137.
1950 Minimum Temperatures and Maximum Flights. Connecticut Valley Hawk Flights—September, 1949. *Bulletin Massachusetts Audubon Society,* 34: 75–80.
1970 A Summary of the 1969 Fall Migration Season, with Special Attention to Eruptions of Various Boreal and Montane Species and an Analysis of Correlations between Wind Flows and Migration. *Audubon Field Notes,* 24 (1): 4–13.
Bailey, S. F.
1967 Fall Hawk Watch at Mt. Peter. *Kingbird,* 17 (3): 129–142.
1969 1968 Hawk Watch at Mt. Peter. *Kingbird,* 19 (4): 200–203.
Barrows, W. B.
1912 Michigan Bird Life. Michigan Agricultural College, East Lansing, Mich.
Baughman, J. L.
1947 A Very Early Notice of Hawk Migrations. *Auk,* 64: 304.
Beaton, R. J.
1951 Hawk Migration at South Mountain. *Atlantic Naturalist,* 6: 166– 168.
Beck, R. A.
1972 Milepost 92: Where Birders Meet. *Virginia Society of Ornithology Newsletter,* 18 (5): 2–3.
Behrend, F. W.
1950a Broad-winged Hawks over Hump Mountain. *Migrant,* 21: 10– 11.
1950b Hawk Migration in Upper East Tennessee. *Migrant,* 21: 70–72.
1951 Fall Migrations of Hawks in 1951. *Migrant,* 22: 53–57.
1952 Fall Migrations of Hawks in 1952. *Migrant,* 23: 62–65.
1953 Hawk Migration—Fall 1953. *Migrant,* 24: 69–73.
1954a Plans for Hawk Watching—1954. *Migrant,* 25: 24–25.
1954b 1954 Fall Migration Count of Hawks. *Migrant,* 25: 69–72.
Bent, A. C.
1937 Life Histories of North American Birds of Prey. Part 1. Bulletin 167. U.S. National Museum, Washington, D.C.
1938 Life Histories of North American Birds of Prey. Part 2. Bulletin 170. U.S. National Museum, Washington, D.C.

Berger, D. D., and H. C. Mueller
  1959 The Bal-Chatri: A Trap for the Birds of Prey. *Bird-Banding,*
        30: 18–26.
Berry, R. B.
  1971 Peregrine Falcon Population Survey, Assateague Island, Mary-
        land, Fall, 1969. *Raptor Research News,* 5: 31–43.
Bihun, A., Jr.
  1967 The Montclair Hawk Lookout Sanctuary 1967 Hawk Watch.
        New Jersey Nature News, 22: 146–147.
  1968 Montclair Hawk Lookout Sanctuary 1968 Hawk Watch. *New
        Jersey Nature News,* 23: 134–135.
  1969 Montclair Hawk Lookout Sanctuary 1969 Hawk Watch. *New
        Jersey Nature News,* 24: 151–153.
  1970 Montclair Hawk Lookout Sanctuary 1970 Hawk Watch. *New
        Jersey Nature News,* 25: 146–148.
  1972 Montclair Bird Club Hawk Watch. *New Jersey Nature News,*
        27 (3): 115–118.
  1973 Montclair Bird Club Hawk Watch Daily Record. *New Jersey
        Nature News,* 28 (2): 85–88.
Bond, J.
  1956 Check-List of Birds of the West Indies. Academy of Natural
        Sciences of Philadelphia, Philadelphia, Pa.
  1968 Thirteenth Supplement to the Check-List of Birds of the West
        Indies (1956). Academy of Natural Sciences of Philadelphia,
        Philadelphia, Pa.
Breck, G. W.
  1959 Operation Hawk-Watch 1959—Final Report on 3 Across-The-
        State Hawk-Watches. Mimeographed Report.
  1960a Operation Hawk-Watch 1960—Results of "3-Week" Watch in
        Upper Montclair, N.J. Mimeographed Report.
  1960b The Montclair Hawk Lookout. *Linnaean News-Letter,* 14,
        No. 5.
Breck, G. W., and R. A. Breck
  1964 The Montclair Hawk Lookout Sanctuary. *New Jersey Nature
        News,* 19: 149–150.
  1965 The Montclair Hawk Lookout Sanctuary 1965 Hawk Watch.
        New Jersey Nature News, 20: 148–149.
  1966 The Montclair Hawk Lookout Sanctuary 1966 Hawk Watch.
        New Jersey Nature News, 21: 164–165.
Breck, R. A.
  1962 Montclair Hawk Lookout Sanctuary. *New Jersey Nature News,*
        17: 138–139.

1963   Montclair Hawk Lookout Sanctuary. *New Jersey Nature News,*
         18: 149–150.
Brett, J. J., and A. C. Nagy
  1973   Feathers in the Wind. Hawk Mountain Sanctuary Association,
         Kempton, Pa.
Broley, C. L.
  1947   Migration and Nesting of Florida Bald Eagles. *Wilson Bulletin,*
         59: 3–20.
Broun, M.
  1935   The Hawk Migration During the Fall of 1934, Along the Kitta-
         tinny Ridge in Pennsylvania. *Auk,* 52: 233–248.
  1936   Three Seasons at Hawk Mountain Sanctuary. Publication 61,
         Emergency Conservation Committee.
  1939   Fall Migration of Hawks at Hawk Mountain, Pennsylvania,
         1934–1938. *Auk,* 56: 429–441.
  1946   Report of the Curator. News Letter to Members No. 15. Hawk
         Mountain Sanctuary Association, Kempton, Pa.
  1947   Hawk Mountain Sanctuary Autumn News from the Curator
         October, November and December, 1946. News Letter to Mem-
         bers No. 16. Hawk Mountain Sanctuary Association, Kempton,
         Pa.
  1949a Hawks Aloft: The Story of Hawk Mountain. Dodd, Mead Co.,
         New York, N.Y.
  1949b Curator's Report 1948. News Letter to Members No. 18. Hawk
         Mountain Sanctuary Association, Kempton, Pa.
  1951   Hawks and the Weather. *Atlantic Naturalist,* 6: 105–112.
  1955   The Curator's Report. News Letter to Members No. 24. Hawk
         Mountain Sanctuary Association, Kempton, Pa.
  1956   Pennsylvania's Bloody Ridges. *Nature Magazine,* 49: 288–292.
  1957   The Curator's Report. News Letter to Members No. 26. Hawk
         Mountain Sanctuary Association, Kempton, Pa.
  1960   Curator's Report—1959. News Letter to Members No. 30. Hawk
         Mountain Sanctuary Association, Kempton, Pa.
  1961   Curator's Report—1960. News Letter to Members No. 31. Hawk
         Mountain Sanctuary Association, Kempton, Pa.
  1963   Hawk Migrations and the Weather. Hawk Mountain Sanctuary
         Association, Kempton, Pa.
  1966   Comments on the Hawk Watch. News Letter to Members No.
         37. Hawk Mountain Sanctuary Association, Kempton, Pa.
Broun, M., and B. V. Goodwin
  1943   Flight-Speeds of Hawks and Crows. *Auk,* 60: 487–492.

Brown, L., and D. Amadon
  1968   Eagles, Hawks and Falcons of the World. Two Volumes. Mc-
          Graw-Hill Book Co., New York, N.Y.
Bull, J.
  1964   Birds of the New York Area. Harper & Row, New York, N.Y.
  1970   Supplement to Birds of the New York Area. *Proc. Linnaean
          Society New York*, 71: 1–54.
Burleigh, T. D.
  1958   Georgia Birds. University Oklahoma Press, Norman, Okla.
Burns, F. L.
  1911   A Monograph of the Broad-winged Hawk (*Buteo platypterus*).
          Wilson Bulletin, 23 (3&4): 143–320.
Cade, T. J.
  1955   Variation of the Common Rough-legged Hawk in America.
          Condor, 57: 313–346.
Cameron, E.
  1964   Hawk Cliff. *Blue Bill*, 11 (3): 31–33.
Carlson, C. W.
  1966   Hawk Watching on the Ridges. *Atlantic Naturalist*, 21: 161–168.
Carpenter, M.
  1949   A Hawk Flight Over Reddish Knob. *Raven*, 20 (9&10): 58–59.
  1960   Observations on Hawk Migrations on the Shenandoah Moun-
          tain, Virginia, 1949–1959. *Raven*, 31 (9&10): 88–90.
Chandler, A. C.
  1914   Modifications and Adaptations to Function in the Feathers of
          Circus hudsonius. *University California Publications Zoology*,
          11 (13): 329–376.
Choate, E. A.
  1972   Spectacular Hawk Flight at Cape May Point, New Jersey, on
          16 October 1970. *Wilson Bulletin*, 84: 340–341.
Choate, E. A., and F. Tilly
  1973   The 1970 Autumn Hawk Count at Cape May Point, New Jersey.
          Science Notes No. 11, New Jersey State Museum, Trenton, N.J.
Christensen, S., B. P. Nielsen, et al.
  1973   Flight Identification of European Raptors. Part 6. Large Fal-
          cons. *British Birds*, 66: 100–114.
Clark, R. J.
  1971   Wing-loading—A Plea for Consistency in Usage. *Auk*, 88: 927–
          928.
Clark, W. S.
  1968   Migration Trapping of Hawks at Cape May, N.J. *EBBA News*,
          31: 112–114.

1969 Migration Trapping of Hawks at Cape May, N.J.—Second Year. *EBBA News,* 32: 69–77.

1970 Migration Trapping of Hawks (and Owls) at Cape May, N.J. —Third Year. *EBBA News,* 33: 181–189.

1971 Migration Trapping of Hawks (and Owls) at Cape May, N.J. —Fourth Year. *EBBA News,* 34: 160–169.

1972 Migration Trapping of Hawks (and Owls) at Cape May, N.J. —Fifth Year. *EBBA News,* 35: 121–131.

1973 Cape May Point Raptor Banding Station—1972 Results. *EBBA News,* 36: 150–165.

Clement, R. C.
1958 Broadwings in Rhode Island. *Narragansett Naturalist,* 1 (4): 118–119.

Clendinning, A. E.
1954 Notes on Fall Hawk Migration—1954. *Cardinal,* 15: 17–24.

Collins, H. H., Jr.
1933 Hawk Slaughter at Drehersville. Bulletin No. 3. Annual Report of the Hawk and Owl Society.

Cone, C. D., Jr.
1961 The Theory of Soaring Flight in Vortex Shells—Part I. *Soaring,* 25 (No. 4).

1962a Thermal Soaring of Birds. *American Scientist,* 50: 180–209.

1962b The Soaring Flight of Birds. *Scientific American,* 260 (April): 130–134, 136, 138, 140.

Cook, A. J.
1893 Birds of Michigan. Bulletin 94. Agricultural Experiment Station, State Agricultural College, Lansing, Michigan.

Cory, C. B.
1909 The Birds of Illinois and Wisconsin. Pub. 131. Field Museum of Natural History, Chicago, Illinois.

Craighead, J. J., and F. C. Craighead, Jr.
1956 Hawks, Owls and Wildlife. Stackpole Co., Harrisburg, Pa.

Cross, A. A.
1927 Observations on Hawk-Banding with Records and Recoveries. *Bulletin Northeastern Bird-Banding Assn.,* 3: 29–33.

Crossan, D. F., and R. A. Stevenson, Jr.
1956 Hawk Migrations Along the Middle Eastern Seaboard, Delaware to North Carolina. *Chat,* 20 (1): 2.

Darrow, H. N.
1963 Direct Autumn Flight-Line from Fire Island, Long Island, to the Coast of Southern New Jersey. *Kingbird,* 13 (1): 4–12.

DeGarmo, W. R.
1953 A Five-Year Study of Hawk Migration. *Redstart,* 20 (3): 39–54.
Dickey, D. R., and A. J. Van Rossem
1938 The Birds of El Salvador. *Field Museum Zoological Series,* 23: 1–609.
Dorst, J.
1961 The Migrations of Birds. Houghton Mifflin Co., Boston, Mass.
Dunbar, R. J.
1950 Hawk Flights in Knox County. *Migrant,* 21: 74.
Dunstan, T. C.
1972 Radio-Tagging Falconiform and Strigiform Birds. *Raptor Research,* 6: 93–102.
Eastwood, E.
1967 Radar Ornithology. Methuen & Co., Ltd., London.
Eaton, E. H.
1910 Birds of New York. Part 1. Memoir 12, New York State Museum, Albany, N.Y.
1914 Birds of New York. Part 2. Memoir 12, New York State Museum, Albany, N.Y.
Edge, R.
1939 News Letter to Members No. 1. Hawk Mountain Sanctuary Association, Kempton, Pa.
1940a News Letter to Members No. 4. Hawk Mountain Sanctuary Association, Kempton, Pa.
1940b News Letter to Members No. 5. Hawk Mountain Sanctuary Association, Kempton, Pa.
Edwards, E. P.
1972 A Field Guide to the Birds of Mexico. Published by the Author, Sweet Briar, Va.
Edwards, J. L.
1939 General Observations of Hawk Migration in New Jersey. *Bulletin* 1 (May): 8–11. Urner Ornithological Club, Newark, N.J.
Eisenmann, E.
1963 Is the Black Vulture Migratory? *Wilson Bulletin,* 75: 244–249.
Elliott, J. J.
1960 Falcon Flights on Long Island. *Kingbird,* 10 (4): 155–157.
Engels, W. L.
1941 Wing Skeleton and Flight of Hawks. *Auk,* 58: 61–69.
Eynon, A. E.
1941 Hawk Migration Routes in the New York City Region. *Proc. Linnaean Society New York,* Nos. 52–53: 113–116.

Fables, D., Jr.
  1955 Annotated List of New Jersey Birds. Urner Ornithological Club, Newark, N.J.

Ferguson, A. L., and H. L. Ferguson
  1922 The Fall Migration of Hawks as Observed at Fishers Island, N.Y. *Auk,* 39: 488–496.

Field, M.
  1970 Hawk-Banding on the Northern Shore of Lake Erie. *Ontario Bird Banding,* 6 (4): 52–69.
  1971 Hawk Cliff Raptor Banding Station First Annual Report, 1971. *Ontario Bird Banding,* 7 (3): 56–75.

Finucane, T. W.
  1956 1955 Fall Migration of Hawks, *Migrant,* 27: 10–12.
  1957 Annual Autumn Hawk Count 1956. *Migrant,* 28: 1–3.
  1958 Annual Autumn Hawk Count 1957. *Migrant,* 29: 1–5.
  1959 Annual Autumn Hawk Count 1958, *Migrant,* 30: 1–5.
  1960 Annual Autumn Hawk Count 1959. *Migrant,* 31: 1–10.
  1961a Annual Autumn Hawk Count 1961 [sic; should be 1960]. *Migrant,* 32: 22–28.
  1961b Annual Autumn Hawk Count 1961. *Migrant,* 32: 57–64.
  1963 Annual Autumn Hawk Count 1962. *Migrant,* 34: 1–7.
  1964 Annual Autumn Hawk Count 1963. *Migrant,* 35: 7–13.
  1965 Annual Autumn Hawk Count 1964. *Migrant,* 36: 1–7.
  1966 Annual Autumn Hawk Count 1965. *Migrant,* 37: 1–3.
  1967 Annual Autumn Hawk Count 1966. *Migrant,* 38: 6–8.
  1968 Annual Autumn Hawk Count, 1967. *Migrant,* 39: 27–29.
  1969 Annual Autumn Hawk Count, 1968. *Migrant,* 40: 28–31.
  1970 Annual Autumn Hawk Count. *Migrant,* 41: 14–18.
  1971 Annual Autumn Hawk Count. *Migrant,* 42: 1–4.
  1972 Annual Autumn Hawk Count. *Migrant,* 43: 35–37, 41.

Fisher, H. I.
  1946 Adaptations and Comparative Anatomy of the Locomotor Apparatus of New World Vultures. *American Midland Naturalist,* 35: 545–727.

Forbes, H. S., and H. B. Forbes
  1927 An Autumn Hawk Flight. *Auk,* 44: 101–102.

Forbush, E. H.
  1927 Birds of Massachusetts and Other New England States. Volume Two. Massachusetts Dept. Agriculture, Boston, Mass.

Foster, G. H.
  1955 Thermal Air Currents and Their Use in Bird-Flight. *British Birds,* XLVIII (6): 241–253.
Frey, E. S.
  1940 Hawk Notes from Sterrett's Gap, Pennsylvania. *Auk,* 57: 247–250.
  1943 Centennial Check-List of the Birds of Cumberland County, Pennsylvania and Her Borders. Published privately, Lemoyne. Pa.
Friedmann, H.
  1950 The Birds of North and Middle America. Bull. 50, United States National Museum, Washington, D.C.
Friedmann, H., L. Griscom, and R. T. Moore
  1950 Distributional Check-List of the Birds of Mexico. Part I. *Pacific Coast Avifauna* No. 29, Cooper Ornithological Club.
Ganier, A. F.
  1951 Migrating Turkey Vultures. *Migrant,* 22: 70.
George, J. C., and A. J. Berger
  1966 Avian Myology. Academic Press, New York, N.Y.
Geyer von Schweppenburg, H. Frieh.
  1963 Zur Terminologie und Theorie der Leitlinie. *J. Ornith.,* 104: 191–204.
Giraud, J. P.
  1844 The Birds of Long Island. Wiley & Putnam, New York, N.Y.
Goldman, H.
  1970 Wings over Brattleboro. *Yankee,* 34 (10): 100–105.
Graham, E. W.
  1972 Swainson's Hawk at Bake Oven Knob, Pennsylvania. *Cassinia,* 53: 43.
Grant, G. S.
  1967 Osprey Migration over Topsail Island, N.C. *Chat,* 31 (4): 96.
Gray, L.
  1961 Banding Sharpshins at Point Pelee. *EBBA News,* 24 (2): 25–26.
Green, J. C.
  1962 1962 Fall Hawk Migration, Duluth. *Flicker,* 34: 121, 124–125.
Griffin, D. R.
  1969 The Physiology and Geophysics of Bird Navigation. *Quart. Review Biology,* 44: 255–276.
  1973 Oriented Bird Migration in or Between Opaque Cloud Layers. *Proc. American Philosophical Society,* 117 (2): 117–141.

Gunn, W. W. H.
   1954   Hints for Hawk Watchers. *Bulletin Federation Ontario Natural-
          ists,* 65: 16–19.
Hackman, C. D.
   1954   A Summary of Hawk Flights over White Marsh, Baltimore
          County, Maryland. *Maryland Birdlife,* 10 (2–3): 19–26.
Hackman, C. D., and C. J. Henny
   1971   Hawk Migration over White Marsh, Maryland. *Chesapeake Sci-
          ence,* 12 (3): 137–141.
Hagar, J. A.
   1937a Hawks at Mount Tom. *Bulletin Massachusetts Audubon So-
          ciety,* 21 (April): 5–8.
   1937b More Hawks at Mount Tom. *Bulletin Massachusetts Audubon
          Society,* 21 (October): 5–8.
Hall, G. A.
   1964   Fall Migration on the Allegheny Front. *Redstart,* 31 (2): 30–53.
Hanley, W.
   1968   Last Call for Hawks. *Narragansett Naturalist,* 10 (1): 2–12.
Hartman, F. A.
   1961   Locomotor Mechanisms of Birds. *Smithsonian Misc. Collections,*
          143 (1): 1–91.
Harwood, M.
   1973   The View from Hawk Mountain. Charles Scribner's Sons, New
          York, N.Y.
Hatch, P. L.
   1892   Notes on the Birds of Minnesota. Geological and Natural His-
          tory Survey, Minneapolis, Minn.
Haugh, J. R.
   1972   A Study of Hawk Migration in Eastern North America. *Search,*
          2 (16): 1–60.
Haugh, J. R., and T. J. Cade
   1966   The Spring Hawk Migration Around the Southeastern Shore of
          Lake Ontario. *Wilson Bulletin,* 78: 88–110.
Hawkes, R. S.
   1958   Hawk Migration in Nova Scotia. *Maine Field Observer,* 3 (11):
          123.
Heintzelman, D. S.
   1961   Kermadec Petrel in Pennsylvania. *Wilson Bulletin,* 73: 262–267.
   1963a Bake Oven Hawk Flights. *Atlantic Naturalist,* 18: 154–158.
   1963b Bake Oven Knob Migration Observations. *Cassinia,* 47: 39–40.

1966   Bake Oven Knob Autumn Hawk Migration Observations (1964 and 1965). *Cassinia,* 49: 33.

1968   Bake Oven Knob Autumn Hawk Migration Observations, 1966 and 1967. *Cassinia,* 50: 26–27.

1969a   The Black Vulture in Pennsylvania. *Pennsylvania Game News,* 40 (5): 17–19.

1969b   Autumn Birds of Bake Oven Knob. *Cassinia,* 51: 11–32.

1970a   Wings Over Hawk Mountain. *National Wildlife,* 8 (5): 22–27.

1970b   Autumn Hawk Watch. *Frontiers,* 35 (1): 16–21.

1970c   Speculation on DDT and Altered Osprey Migrations. *Raptor Research News,* 4: 120–124.

1970d   Bake Oven Knob Autumn Hawk Migration Observations (1957 and 1969). *Cassinia,* 52: 37.

1970e   The Hawks of New Jersey. Bulletin 13. New Jersey State Museum, Trenton, N.J.

1970f   An Intensive Inter-Specific Aerial Display between a Sharp-shinned Hawk and a Sparrow Hawk. *Cassinia,* 52: 39.

1972a   The 1971 Autumn Hawkwatch at Catfish Fire Tower, New Jersey. *New Jersey Nature News,* 27 (1): 19–21.

1972b   Some Autumn Bird Records from the Catfish Fire Tower, New Jersey. *Science Notes* No. 6, New Jersey State Museum, Trenton, N.J.

1972c   The Importance of Record Keeping when Watching Hawk Migrations. *California Condor,* 7 (4): 6–7.

1972d   A Guide to Northeastern Hawk Watching. Privately published, Lambertville, N.J.

1972e   Hawks Across New Jersey. *New Jersey Nature News,* 27: 100–104.

1972f   Speculation on the Possible Origin of some Sharp-shinned and Red-tailed Hawk Flights along the Kittatinny Ridge—Autumn 1971. *Science Notes* No. 10, New Jersey State Museum, Trenton, N.J.

1973   The 1972 Autumn Hawk Count at Tott's Gap, Pennsylvania. *Science Notes* No. 12, New Jersey State Museum, Trenton, N.J.

Heintzelman, D. S., and T. V. Armentano

1964   Autumn Bird Migration at Bake Oven Knob, Pa. *Cassinia,* 48: 2–18.

Heintzelman, D. S., and R. MacClay

1971   An Extraordinary Autumn Migration of White-breasted Nuthatches. *Wilson Bulletin,* 83: 129–131.

1972 The 1970 and 1971 Autumn Hawk Counts at Bake Oven Knob, Pennsylvania. *Cassinia*, 53: 3–23.

1973 The 1972 Autumn Hawk Count at Bake Oven Knob, Pennsylvania. *Cassinia*, 54: 3–9.

Hellmayr, C. E., and B. Conover

1949 Catalogue of Birds of the Americas and the Adjacent Islands. Part I, No. 4. Field Museum Natural History, *Zool. Series*, 13: 1–358.

Herndon, L. R.

1949 Golden Eagle at Hump Mountain, Tenn.-N.C. *Migrant*, 20: 17.

Hickey, J. J. (Ed.)

1969 Peregrine Falcon Populations/Their Biology and Decline. University Wisconsin Press, Madison, Wisc.

Hicks, D. L., D. T. Rogers, Jr., and G. I. Child

1966 Autumnal Hawk Migration Through Panama. *Bird-Banding*, 37: 121–123.

Hill, R. K.

1957 Hawk Watching. *New Hampshire Bird News*, 10 (3): 64–67.

Hofslund, P. B.

1954 The Hawk Pass at Duluth, Minnesota. *Wilson Bulletin*, 66: 224.

1962 The Duluth Hawk Flyway—1951–1961. *Flicker*, 34: 88–92.

1966 Hawk Migration Over the Western Tip of Lake Superior. *Wilson Bulletin*, 78: 79–87.

Hopkins, D. A., and G. S. Mersereau

1971 Fall Hawk Migration 1971. Privately published.

1972 Hawk Migration 1972. Privately published.

Howell, A. H.

1932 Florida Bird Life. Coward-McCann, Inc., New York, N.Y.

Hurley, G.

1970 Fall Hawk Migration along Peters Mountain in Monroe County, West Virginia. *Redstart*, 37 (3): 82–86.

Imhof, T. A.

1962 Alabama Birds. University Alabama Press, University, Ala.

Johnson, N. K., and H. J. Peeters

1963 The Systematic Position of Certain Hawks in the Genus *Buteo*. *Auk*, 80: 417–446.

Johnson, W. M.

1950 Migrating Hawks in the Blue Ridge. *Migrant*, 21: 72–74.

Jung, C. S.

1935 Migration of Hawks in Wisconsin. *Wilson Bulletin*, 47: 75–76.

1964 Weather Conditions Affecting Hawk Migrations. *Lore,* 14 (4): 134–142.

Keith, L. B.
1963 Wildlife's Ten-Year Cycle. University Wisconsin Press, Madison, Wisc.

Kleen, V. M.
1969 Banding North American Migrants in Panama. *EBBA News,* 32: 170–173.

Kleiman, J. P.
1966 Migration of Rough-legged Hawks over Lake Erie. *Wilson Bulletin,* 78: 122.

Koebel, T.
1970 Bearfort Tower Lookout. New Jersey Highland Hawk Observation Notes—1968–1969. *Urner Field Observer,* 12 (1): 9–21.

Koford, C. B.
1953 The California Condor. Research Report No. 4, National Audubon Society, New York, N.Y.

Lack, D.
1954 The Natural Regulation of Animal Numbers. Oxford University Press, London.

Land, H. C.
1970 Birds of Guatemala. Livingston Publishing Co., Wynnewood, Pa.

Lang, E. B.
1943 Hawk Watching in the Watchungs. *Audubon Magazine,* 45: 346–351.

Leck, C. F.
1972 Wave Phenomena of Land Migrants at Island Beach State Park, New Jersey. *Bird-Banding,* 43: 20–25.

Leopold, N. F.
1963 Checklist of Birds of Puerto Rico and the Virgin Islands. Bulletin 168, Agric. Expt. Station, University Puerto Rico, Rio Piedras, Puerto Rico.

Lincoln, F. C.
1936 Recoveries of Banded Birds of Prey. *Bird-Banding,* 7: 38–45.

Lowery, G. H., Jr.
1960 Louisiana Birds. Revised second edition. Louisiana State University Press, Baton Rouge, La.

Lowery, G. H., Jr., and R. J. Newman
1954 The Birds of the Gulf of Mexico. Fishery Bulletin 89, United States Fish and Wildlife Service, Washington, D.C. Pp. 519–540.

McIlhenny, E. A.
  1939   An Unusual Migration of Broad-winged Hawks. *Auk,* 56: 182–
         183.
McMillan, I.
  1968   Man and the California Condor. E. P. Dutton & Co., New York,
         N.Y.
Meng, H.
  1963   Radio Controlled Hawk Trap. *EBBA News,* 26: 185–188.
Mengel, R. M.
  1965   The Birds of Kentucky. Ornithological Monographs No. 3, Amer-
         ican Ornithologists' Union.
Merriam, C. H.
  1877   A Review of the Birds of Connecticut with Remarks on their
         Habits. *Trans. Connecticut Academy,* 4: 1–150.
Merriam, R. D.
  1953   Detroit Region Hawk Observations, Fall 1953.
  1954a Hawk Count, Fall 1953. *Bird Survey of the Detroit Region,*
         1953.
  1954b Detroit Region Hawk Observations, Fall 1954.
  1956   Hawk Migration, Fall 1954. *Bird Survey of the Detroit Region,*
         1954.
Miller, A. D.
  1952   Hawk Count, 1951. *Bird Survey of the Detroit Region,* 1951.
Miller, B. L.
  1941   Lehigh County Pennsylvania Geology and Geography. Bulletin
         C39, Fourth series, Pennsylvania Geological Survey, Harris-
         burg, Pa.
Miller, A. H., I. I. McMillan, and E. McMillan
  1965   The Current Status and Welfare of the California Condor. Re-
         search Report No. 6, National Audubon Society, New York,
         N.Y.
Mills, E., and L. Mills
  1971   Hook Mountain, N.Y. Hawk Watch. *California Condor,* 6 (4):
         12.
Mohr, C. E.
  1969   Our Handsome Birds of Prey: Hawks and Eagles. *Delaware
         Conservationist,* 13 (4): 3–13.
Monroe, B. L., Jr.
  1968   A Distributional Survey of the Birds of Honduras. Ornitholog-
         ical Monographs No. 7, American Ornithologists' Union.

Mueller, H. C., and D. D. Berger

1961 Weather and Fall Migration of Hawks at Cedar Grove, Wisconsin. *Wilson Bulletin*, 73: 171–192.

1965 A Summer Movement of Broad-winged Hawks. *Wilson Bulletin*, 77: 83–84.

1966 Analyses of Weight and Fat Variations in Transient Swainson's Thrushes. *Bird-Banding*, 37: 83–112.

1967a Wind Drift, Leading Lines, and Diurnal Migrations. *Wilson Bulletin*, 79: 50–63.

1967b Fall Migration of Sharp-shinned Hawks. *Wilson Bulletin*, 79: 397–415.

1968 Sex Ratios and Measurements of Migrant Goshawks. *Auk*, 85: 431–436.

1973 The Daily Rhythm of Hawk Migration at Cedar Grove, Wisconsin. *Auk*, 90: 591–596.

Murray, B. G., Jr.

1964 A Review of Sharp-shinned Hawk Migration along the Northeastern Coast of the United States. *Wilson Bulletin*, 76: 257–264.

1969 Sharp-shinned Hawk Migration in the Northeastern United States. *Wilson Bulletin*, 81: 119–120.

Murton, R. K., and E. N. Wright

1968 The Problems of Birds as Pests. Academic Press, New York, N.Y.

Nagy, A. C.

1967 Curator's Report—1966. News Letter to Members No. 39. Hawk Mountain Sanctuary Association, Kempton, Pa.

1970 Curator's Report. News Letter to Members No. 42. Hawk Mountain Sanctuary Association, Kempton, Pa.

1972 1971 Curator's Report. News Letter to Members No. 44. Hawk Mountain Sanctuary Association, Kempton, Pa.

1973 1972 Curator's Report. News Letter to Members No. 45. Hawk Mountain Sanctuary Association, Kempton, Pa.

Oberholser, H. C.

1938 The Bird Life of Louisiana. Bulletin 28. Louisiana Department of Conservation, New Orleans, La.

Odom, T.

1966 Summary of Broad-winged Hawk Flights Across Tennessee from 1951 through 1964. *J. Tennessee Academy Science*, 41: 95–96.

Olendorff, R. R., and S. E. Olendorff

1968 An Extensive Bibliography on Falconry, Eagles, Hawks, Falcons and other Diurnal Birds of Prey. Part 1. Falconry and Eagles. Published privately, Fort Collins, Colorado.

1969   An Extensive Bibliography on Falconry, Eagles, Hawks, Falcons
       and other Diurnal Birds of Prey. Part 2. Hawks and Miscellan-
       eous. Published privately, Fort Collins, Colorado.

1970   An Extensive Bibliography on Falconry, Eagles, Hawks, Falcons
       and other Diurnal Birds of Prey. Part 3. Falcons and Ospreys.
       Published privately, Fort Collins, Colorado.

Parkes, T.
1957   Hawk Migration. *Chat,* 21 (4): 88–89.

Pennycuick, C. J.
1973   The Soaring Flight of Vultures. *Scientific American,* 229 (6):
       102–109.

Perkins, J. P.
1964   17 Flyways over the Great Lakes. Part 1. *Audubon Magazine,*
       66 (5): 294–299.

Peters, H. S., and T. D. Burleigh
1951   The Birds of Newfoundland. Houghton Mifflin Co., Boston,
       Mass.

Peterson, R. T.
1947   A Field Guide to the Birds. Second Revised and Enlarged Edi-
       tion. Houghton Mifflin Co., Boston, Mass.

1960   A Field Guide to the Birds of Texas. Houghton Mifflin Co., Bos-
       ton, Mass.

1966   Tribute to Maurice and Irma Broun. News Letter to Members
       No. 38. Hawk Mountain Sanctuary Association, Kempton, Pa.

Peterson, R. T., and E. L. Chalif
1973   A Field Guide to Mexican Birds and Adjacent Central America.
       Houghton Mifflin Co., Boston, Mass.

Pettingill, O. S., Jr.
1962   Hawk Migrations Around the Great Lakes. *Audubon Magazine,*
       64 (1): 44–45, 49.

Pettingill, O. S., Jr., and S. F. Hoyt
1963   Enjoying Birds in Upstate New York. Laboratory of Ornithol-
       ogy, Cornell University, Ithaca, N.Y.

Poole, E. L.
1932   A Survey of the Mammals of Berks County, Pennsylvania. Bul-
       letin 13, Reading Public Museum and Art Gallery, Reading, Pa.

1934   The Hawk Migration Along the Kittatinny Ridge in Pennsyl-
       vania. *Auk,* 51: 17–20.

1938   Weights and Wing Areas in North American Birds. *Auk,* 55:
       511–517.

Pough, R. H.
1932 Wholesale Killing of Hawks in Pennsylvania. *Bird-Lore,* 34 (6): 429–430.
1936 Pennsylvania and the Hawk Problem. *Pennsylvania Game News,* 7 (4): 8, 23.
1951 Audubon Water Bird Guide: Water, Game and Large Land Birds. Doubleday & Co., New York, N.Y.

Powell, S. E.
1957 Notes on a Migration of Broad-winged Hawks. *Maine Field Observer,* 2 (10).

Pratt, D.
1967 Observation of Broad-winged Hawk Migration at Table Rock, N.C. *Chat,* 31 (4): 95.

Raspet, A.
1960 Biophysics of Bird Flight. *Science,* 132 (3421): 191–199.

Recher, H. F., and J. T. Recher
1966 A Contribution to the Knowledge of the Avifauna of the Sierra de Luquillo, Puerto Rico. *Caribbean J. Science,* 6 (3–4): 151–161.

Redmond, E. C., and R. A. Breck
1961 Montclair Hawk Lookout Sanctuary. *New Jersey Nature News,* 16: 118, 123.

Reese, J. G.
1973 Bald Eagle Migration along the Upper Mississippi River in Minnesota. *Loon,* 45 (1): 22–23.

Robbins, C. S.
1950 Hawks over Maryland, Fall of 1949. *Maryland Birdlife,* 6 (1): 2–11.
1956 Hawk Watch. *Atlantic Naturalist,* 11 (5): 208–217.
1966 A Guide to Field Identification/Birds of North America. Golden Press, New York, N.Y.

Roberts, T. S.
1932 The Birds of Minnesota. Vol. 1. University Minnesota Press, Minneapolis, Minn.

Robertson, W. B., Jr.
1970 Florida Region. *Audubon Field Notes,* 24 (1): 34–35.
1971 Florida Region. *American Birds,* 25: 46–47.
1972 Florida Region. *American Birds,* 26: 51.

Robertson, W. B., Jr., and J. C. Ogden
1968 Florida Region. *Audubon Field Notes,* 22: 25–31.

Rogers, R.
   1971   Highlands Audubon Society Mt. Peter, N.Y., Hawk Watch,
          1971. *California Condor*, 6 (4): 13.
   1972   Hawk Watch '72. *Highlands Audubon Society Newsletter*, 2
          (12): 6–7.
Rusling, W. J.
   1936   The Study of the Habits of Diurnal Migrants, as Related to
          Weather and Land Masses during the Fall Migration on the
          Atlantic Coast, with Particular Reference to the Hawk Flights
          of the Cape Charles (Virginia) Region. Unpublished Manu-
          script.
Russell, S. M.
   1964   A Distributional Study of the Birds of British Honduras. Orni-
          thological Monographs No. 1, American Ornithologists' Union.
Sage, J. H., L. B. Bishop, and W. P. Bliss
   1913   The Birds of Connecticut. Bulletin 20, Connecticut Geology and
          Natural History Survey, Hartford, Conn.
Saville, D. B. O.
   1957   Adaptive Evolution in the Avian Wing. *Evolution*, 11 (2): 212–
          224.
Sharadin, R.
   1972   1971 Hawk Migration. News Letter to Members No. 44. Hawk
          Mountain Sanctuary Association, Kempton, Pa.
   1973   1972 Hawk Migration. News Letter to Members No. 45. Hawk
          Mountain Sanctuary Association, Kempton, Pa.
Sheldon, W.
   1965   Hawk Migration in Michigan and the Straits of Mackinac. *Jack-
          Pine Warbler*, 43 (2): 79–83.
Shreve, A.
   1970   Broadwings Over Kanawha County, September 1969. *Redstart*,
          37 (4): 115–116.
Simpson, T. W.
   1952   A Preliminary Note on the 1952 Hawk Migration Project. *Chat*,
          16 (4): 92.
   1954   The Status of Migratory Hawks in the Carolinas. *Chat*, 18 (1):
          15–21.
Skutch, A. F.
   1945   The Migration of Swainson's and Broad-winged Hawks through
          Costa Rica. *Northwest Science*, 19 (4): 80–89.

Slud, P.
   1964 The Birds of Costa Rica. Bulletin 128, American Museum of Natural History, New York, N.Y. Pp. 1–430.

Smith, N.
   1973 Spectacular Buteo Migration over Panama Canal Zone, October, 1972. *American Birds,* 27 (1): 3–5.

Smithe, F. B.
   1966 The Birds of Tikal. Natural History Press, Garden City, N.Y.

Southern, W. E.
   1964 Additional Observations on Winter Bald Eagle Populations: Including Remarks on Biotelemetry Techniques and Immature Plumage. *Wilson Bulletin,* 76: 121–137.

Spofford, W. R.
   1949 Fall Flight of Hawks at Fall Creek Falls State Park. *Migrant,* 20 (1): 16.
   1969 Hawk Mountain Counts as Population Indices in Northeastern America. *In* Peregrine Falcon Populations (J. J. Hickey, ed.), University Wisconsin Press, Madison, Wisc.
   1971 The Breeding Status of the Golden Eagle in the Appalachians. *American Birds,* 25 (1): 3–7.

Sprunt, A., Jr.
   1954 Florida Bird Life. Coward-McCann, New York, N.Y.

Sprunt, A., IV
   1969 Population Trends of the Bald Eagle in North America. *In* Peregrine Falcon Populations (J. J. Hickey, ed.), University Wisconsin Press, Madison, Wis.

Sprunt, A., Jr., and E. B. Chamberlain
   1970 South Carolina Bird Life. Revised edition, with a supplement. University South Carolina Press, Columbia, S.C.

Squires, W. A.
   1952 The Birds of New Brunswick. Monograph Series No. 4. New Brunswick Museum, Saint John, N.B.

Stearns, E. I.
   1948a A Study of the Migration of the Broad-winged Hawk through New Jersey. *Urner Field Observer,* 3 (3&4): 1–4.
   1948b Blawking: The Study of Hawks in Flight from a Blimp. *Urner Field Observer,* 3 (5&6): 2–9.
   1949 The Study of Hawks in Flight from a Blimp. *Wilson Bulletin,* 61: 110.

Stewart, R. E., and C. S. Robbins
    1958   Birds of Maryland and the District of Columbia. *North American Fauna* No. 62, Washington, D.C.
Stone, W.
    1922   Hawk Flights at Cape May Point, N.J. *Auk,* 39: 567–568.
    1937   Bird Studies at Old Cape May. Volume 1. Delaware Valley Ornithological Club, Philadelphia, Pa.
Storer, J. H.
    1948   The Flight of Birds. Bulletin 28, Cranbrook Institute of Science, Bloomfield Hills, Mich.
Storer, R. W.
    1955   Weight, Wing Area, and Skeletal Proportions in Three Accipiters. *Acta XI Congress International Ornithology:* 287–290.
Sundquist, K.
    1973   Hawk Watch Duluth 1972 Summary. *Hawk Ridge Nature Reserve Newsletter,* 1: 1–9.
Sutton, G. M.
    1928   Notes on a Collection of Hawks from Schuylkill County, Pennsylvania. *Wilson Bulletin,* 40: 84–95, 193–194.
Sutton, W. D.
    1956   Hawk Cliff History. *Cardinal,* 22: 7–12.
Tabb, E. C.
    1973   A Study of Wintering Broad-winged Hawks in Southwestern Florida, 1968–1973. *EBBA News,* 36 (Supple.): 11–29.
Taylor, J. W.
    1970   President's Message. News Letter to Members No. 42. Hawk Mountain Sanctuary Association, Kempton, Pa.
    1971   President's Message. News Letter to Members No. 43. Hawk Mountain Sanctuary Association, Kempton, Pa.
Thomson, A. L. (Ed.)
    1964   A New Dictionary of Birds. McGraw-Hill Book Co., New York, N.Y.
Thomas, S.
    1971a Hook Mountain, New York, Hawk Watch. *California Condor,* 6 (3): 13.
    1971b Hook Mt. Hawk Watch—Autumn, 1971. *California Condor,* 6 (5): 10–12.
    1973   Annual Report Hook Mountain Hawk Watch. Mimeographed report.

Tilly, F.
  1972a The 1971 Raccoon Ridge Autumn Hawkwatch, *New Jersey Nature News,* 27 (1): 22–28.
  1972b Raccoon Ridge Daily Hawk Counts—Autumn 1971. *Science Notes* No. 7, New Jersey State Museum, Trenton, N.J.
  1973 The 1972 Autumn Hawk Count at Raccoon Ridge, Warren County, New Jersey. *Science Notes* No. 13, New Jersey State Museum, Trenton, N.J.
Todd, W. E. C.
  1963 Birds of the Labrador Peninsula and Adjacent Areas. University Toronto Press, Toronto.
Trowbridge, C. C.
  1895 Hawk Flights in Connecticut. *Auk,* 12: 259–270.
Tufts, R. W.
  1962 The Birds of Nova Scotia. Nova Scotia Museum, Halifax.
Van Eseltine, W. P.
  1952 A Hawk Migration Project for the Carolinas. *Chat,* 16 (1): 6.
Van Tyne, J., and G. M. Sutton
  1937 The Birds of Brewster County, Texas. *Misc. Publications* No. 37, Museum of Zoology, University Michigan, Ann Arbor, Mich.
von Lengerke, J.
  1908 Migration of Hawks. *Auk,* 25: 315–316.
Ward, C. J.
  1958 Hawk Flights at Jones Beach, Long Island. *Kingbird,* 8 (2): 42–43.
  1960a Hawk Flights at Jones Beach. *Linnaean News-Letter,* 14, No. 2.
  1960b Hawk Migrations at Jones Beach. *Kingbird,* 10 (4): 157–159.
  1963 Fall Hawk Migrations of Region 10. *Kingbird,* 13 (1): 22–23.
Ward, F. P., and R. B. Berry
  1972 Autumn Migrations of Peregrine Falcons on Assateague Island, 1970–71. *J. Wildlife Management,* 36 (2): 484–492.
Wauer, R. H.
  1973 Birds of Big Bend National Park and Vicinity. University Texas Press, Austin, Texas.
Wellman, C.
  1957 September Hawk Flight. *New Hampshire Bird News,* 10 (3): 77.
Wetmore, A.
  1965 The Birds of the Republic of Panama. Part 1. Smithsonian Miscellaneous Collections, Vol. 150, Washington, D.C.

Wetzel, F. W.
   1969   The Hawk Season. News Letter to Members No. 41. Hawk
          Mountain Sanctuary Association, Kempton, Pa.
Wolfe, L. R.
   1956   Check-List of the Birds of Texas. Intelligencer Printing Co.,
          Lancaster, Pa.
Wood, M.
   1967   Birds of Pennsylvania. Pennsylvania Agricultural Experiment
          Station, University Park, Pa.
Wood, N. A.
   1951   The Birds of Michigan. Misc. Publications No. 75. Museum of
          Zoology, University Michigan, Ann Arbor, Mich.
Worth, C. B.
   1936   Summary and Analysis of Some Records of Banded Ospreys.
          Bird-Banding, 7: 156–160.
Wray, D. L., and H. T. Davis
   1959   Birds of North Carolina. Revised ed. Bynum Printing Co.,
          Raleigh, N.C.

Index

Index by Lisa McGaw

About the Author

Donald S. Heintzelman, currently a wildlife consultant, a lecturer, and a writer, has been an Associate Curator of Natural Science at the William Penn Memorial Museum and was for some years Curator of Ornithology at the New Jersey State Museum. He is the author of *The Hawks of New Jersey* and *A Guide to Northeastern Hawk Watching* as well as of many articles and notes in journals of national and regional ornithology and conservation periodicals. He is an Audubon Wildlife Film lecturer, and his photographs have appeared as illustrations in many magazine articles and books. He is preparing a volume entitled *Hawks and Owls of North America*.